Soviet Oil

The Move Offshore

Soviet Oil

The Move Offshore

Stephen Lewarne

Routledge
Taylor & Francis Group
LONDON AND NEW YORK

First published 1988 by Westview Press, Inc.

Published 2019 by Routledge
52 Vanderbilt Avenue, New York, NY 10017
2 Park Square, Milton Park, Abingdon, Oxon OX14 4RN

Routledge is an imprint of the Taylor & Francis Group, an informa business

Library of Congress Cataloging-in-Publication Data
Lewarne, Stephen.
Soviet oil: the move offshore/by Stephen Lewarne.
p. cm.—(Westview special studies on the Soviet Union and Eastern Europe)
Bibliography: p.
Includes index.
ISBN 0-8133-7641-6
1. Offshore oil industry—Soviet Union. 2. Offshore oil industry—Arctic Ocean. I. Title. II. Series.
HD9575.S652L48 1988
333.8'232'0947—dc19 **88-10642**
CIP

ISBN 13: 978-0-367-28832-7 (hbk)

ISBN 13: 978-0-367-30378-5 (pbk)

To David Campbell, Peter Gould,
Richard LePage and Kathleen Reyno

Contents

Maps

Tables

Acknowledgments

My research on this work began while I was at the Institute of Soviet and East European Studies, Carleton University, in Ottawa, Canada, from 1985 to 1987, and I am most indebted to the institute and its members. Particularly, I would like to express my deepest thanks and admiration for Prof. Carl McMillan, whose patience and guidance have seen me through nearly eight years of Soviet affairs. It was his idea that I undertake a study on the Soviet offshore oil industry. His research guidance and comments on several early drafts were crucial to the study's development. I would also like to thank Prof. Larry Black, Prof. Richard Carson, Donna Harper, Carl Reid, Peter Gould, and David Campbell--all members of the institute.

Special thanks must be given to John Hannigan, who obtained several crucial documents and studies for me from Circumpolar Affairs of the Canadian Government, and to Arpad Abonyi of Investment Canada. My thanks also go to Sergei Vecherskii and Prof. Bogomosov of the Economics Department of Leningrad (Zhdanov) University, both of whom assisted my research efforts while I was on a special study term in the Soviet Union in 1986.

Maureen Korp prepared the final text edit and Kathleen Hutnik a preliminary edit. Their comments and corrections were needed and appreciated. Peter Gould and Lt. Michael Geskin reviewed the transliteration in the book and provided me with useful advice. John Hollingsworth of Indiana University's Department of Geography and Barbara McCuaig of Interact Ltd., Ottawa, did the cartography work.

At Indiana University I want to thank Prof. Robert Campbell for his help in the final publishing stages of the book, Prof. Christopher Waller for his support, Dorene Cornwell for computer assistance, and Karen Niggle of the Indiana Russian and East European Institute.

Michael Kelly, Patrick Caragata, Alan Darrise, Kathleen Reyno, Ken Hart, Chummer Farina, Catherine Disher and

Richard LePage all read drafts of the manuscript and offered useful advice.

Susan Laury was especially helpful in the final few months of preparing the manuscript for publication, as were Carlos Pinerua and Dean Dudley. Thanks must also be expressed to my mother and father and Andrew Lewarne for their support.

Finally, I would like to thank the Mellon Foundation, the Indiana Russian and East European Institute, the Canadian Department of National Defence, and the Office of Associate Dean Timothy Londergan of Indiana University's Graduate School. All provided financial support for the study.

I am, of course, responsible for any errors or omissions.

Stephen M. Lewarne
Indiana University
Bloomington, Indiana

1

The Importance of Soviet Petroleum and the Role of Offshore Oil: An Introduction

The Soviet Union has elicited close scrutiny in the West over its role as both an oil producer and exporter. This scrutiny has become more pronounced over the last decade when it became apparent that the Soviet Union was truly an oil "giant". It had become the largest producer of oil in the world by the mid-1970s, a position previously held by the United States. Further, when the Soviet Union in 1979 surpassed Iran, becoming second only to Saudi Arabia in oil exports, it became clear to analysts that petroleum was a key source of the Soviet Union's economic strength and was crucial to the Soviet Union's presence in the global economy. The production of oil for the Soviet Union, therefore, is more than just an economic necessity, it is a source of prestige and an instrument of political power.

Over the last decade the Soviet Union has experienced problems in its petroleum industry. Among these are, notably, the decreasing rates of growth both for petroleum production and the discovery of new petroleum reserves. The annual rate of growth in petroleum production has fallen from an average of 9% between 1960 and 1970 to 4% between 1975 and 1980, and to -0.25% between 1980 and 1985. Particularly disturbing to the Soviets was the absolute decline between 1983 and 1985 when production went from

616 million tons a year (mty) to 596 mty. This was increased to 611 mty in 1986 and is estimated at 624 mty in 1987. The production increases in these two years are, however, deceptive because they were produced at an increasing cost. Most of the increase is attributable to more discipline at the work sites than to long-term factors such as increased exploratory drilling and technical innovation. What has not been addressed is the crucial decline in the reserves-to-production ratio (the amount of known explored reserves as a share of oil production) of -3.9% between 1970 and 1984. It is unlikely that production figures like those of 1987 will continue if these problems are not addressed. Because petroleum is the Soviet Union's chief earner of hard currency, these problems are that much more acute.

Soviet petroleum is significant in five ways. First, if petroleum production begins to fall off to the extent that it cannot satisfy domestic needs, then serious bottlenecks will begin to form in the economy until other energy sources can be drawn upon. Second, petroleum has been and still is the Soviet Union's chief earner of hard currency. For quite a while it has been the preoccupation of Soviet planners to raise the quality of Soviet manufactured goods and increase their salability in hard currency markets. Yet to date for a number of reasons this has not occurred. The consequence is that petroleum is really the principal "tool" available to Soviet planners when they wish to reduce the country's trade deficit or increase earnings to pay for hard currency imports. Third, in the post-OPEC world, oil still equates with power. Security of reserves can allow a country considerable leverage in its economic and political dealings with both its allies and its enemies. In the case of the Soviet Union, a surplus of oil means it can remain aloof at times from the Middle East conflict when the conflict centers around oil. Fourth, Soviet oil is significant as an instrument of Soviet power in East Europe. Fifth, if the reserves-to-production ratio begins to seriously decline, then the Soviet Union and those who rely upon its petroleum for their energy needs may be forced to turn to the Middle East for petroleum. Such concentration could increase tension in a region already prone to conflict.

The Soviet Union has tended to benefit from being an oil giant, both in economic and political terms. It has, however, attributed much of its success to oil and oil alone. Oddly, just when the Soviet Union had become the world's largest producer and second-largest exporter of oil, the Soviet Union began to experience problems in its petroleum industry.

Soviet planners under Mikhail Gorbachev are now considering exploring and developing the country's offshore waters in hope of providing at least a partial solution to these problems. With an unstable rate of growth in petroleum production since 1984 and the inadequate exploratory drilling of the past fifteen years, the Soviet Union is now faced with a decline in the reserves-to-production ratio. Further, over the long term, its major source of hard currency is jeopardized.

Are Soviet planners concerned with the extent to which offshore oil can alleviate these problems? The answer depends.

If their concern is to know if petroleum from the offshore regions can provide a solution to current production problems of the Soviet petroleum industry, then the answer is "no". At the present time, the offshore regions of the Soviet Union produce around 10 mty, or 200 thousand barrels a day (tbd), which is about 1.5% of total Soviet petroleum production. On the other hand, if their concern is to know if the offshore regions have the potential to alleviate significantly the problem of a declining reserves-to-production ratio, then the answer is most definitely "yes".

However, the term "potential" is used cautiously. What is meant is that the geological potential of the offshore waters appears to be substantial, and few experts, both Soviet and Western, doubt the potential reserve capacity of the offshore waters. Some argue whether this potential capacity amounts to 98 billion barrels (bbl) or 151 bbl, but even pessimistic estimates place the reserve capacity among the largest in the world.[1] According to the East German journal *Die Wirtschaft*, Soviet waters contain about one-third of the country's ultimate recoverable reserves. The inland seas alone are said to have recoverable reserves of nearly 39bbl.[2]

Given that the offshore reserves are believed to exist in substantial quantities, this study will address the questions of the practical ability of these reserves to alleviate significantly the problems of a declining reserves-to-production ratio and to insure long-term petroleum production. To this end several issues are raised.

To what extent will problems in the onshore petroleum industry affect the decision to go offshore? That is, how crucial is petroleum to the Soviet Union in the years ahead? Is the Soviet Union prepared to spend a considerable amount of money and take severe risks to develop the offshore waters?

How is the Soviet Union to assess, explore and develop this new potential? In most of the petroleum-producing regions of the Soviet Union, the lack of adequate technology and the remoteness of the region are the major impediments. The ultimate problem, however, "...will be one of technology, because even if the most remote basin is explored successfully by geophysical means, it will take a modern, advanced drilling technology to develop the resource potential at the basin."[3]

The primary impediment, therefore, is a technological one. In this regard, the Soviet Union must know its technological requirements; *e.g.*, what kind of offshore equipment is necessary to undertake seismological surveys, exploratory drilling and production? How significant is the difference between the more temperate offshore region of the Caspian Sea and the Arctic conditions of the Barents Sea and the Sea of Okhotsk?

The Soviet Union's domestic industry is not currently capable of meeting these requirements. As a result it has enlisted the support of Western firms and governments. The participation of Western countries, and in many cases NATO members, entails a number of problems.

First, how will the Soviet Union pay for this technology? One of the main points stressed in this study, is the paramount importance of petroleum in Soviet export earnings. Importing foreign technology, however, means that more hard currency must be paid out in order to acquire the equipment. Since over two-thirds of the Soviet Union's hard currency earnings are derived from oil sales, with every purchase of Western equipment, more oil must be produced to cover the cost of acquisition.

Second, what are the security considerations if Western partners are allowed to participate in the development of the offshore reserves? With the import of foreign technology and, perhaps, Western engineers and technicians to assist in the proper use of the equipment, there is a possibility of Western influence in a region traditionally off limits to foreigners.

Third, if Western firms or governments become involved in the development of the shelf, even at an arm's length via the licensing of technology, how will political conflicts between the United States and the Soviet Union--such as the Soviet invasion of Afghanistan or American sanctions of the Soviet-West Europe gas pipeline--affect business practices between the Soviet Union and its Western partners?

Fourth, to what extent will Soviet planners attempt to reduce systemic barriers inherent in the Soviet economy to facilitate the development of the offshore waters? The centrally planned structure of the Soviet economy and the monopoly over foreign trade by the foreign trade organizations (FTOs) more often than not slow down negotiations. This leads to frustration on the part of the Western partner.

Finally, what of the potential territorial disputes? The exploration and development of the offshore region involves in many instances conflict over boundaries with the Soviet Union's marine neighbors. While most of the regions around the Soviet Union have a potential for dispute, there are three regions of immediate consideration--the waters between the island of Sakhalin and Japan; the Barents Sea and Norway; and, the northern polar ice cap and Canada.

While the dispute over ownership of the island of Sakhalin was settled in 1952, with Japan renouncing claims to the region south of Karafuto, the dispute over possession of the Kuril islands still exists. The marine border between Sakhalin and the Kurils has not yet been formally resolved between the Soviet Union and Japan.

In the Barents Sea, there is what is known as the "Grey Zone" an area between Norway and the Soviet Union. The rights to mineral use have never been decided. As the Norwegians move farther and farther north towards Spitsbergen and into the Murmansk basin in an effort to expand their offshore oil industry, this area will be the most hotly debated. A similar dispute already exists between the two countries concerning Spitsbergen island and the waters around it.

Although there is no dispute with Canada yet, the potential certainly exists, because Canada has decided recently to explore more actively its Arctic waters. Canada may find itself having to settle boundaries with the Soviet Union. The region of potential dispute between the two countries is the Lomonosov ridge extending under the northern polar ice cap.

The island of Gotland in the Baltic Sea, where oil has been discovered, is a more innocuous dispute. Although the Soviet Union lays claim to half the island, Sweden claims an area encompassing the island, a distance halfway between Sweden and the Soviet Union.

There are other areas of potential concern. For example, the Bering Sea between the Soviet Union and the United States is the longest shared marine border in the world; however, the Convention of 1869, ceding Alaska to the United States, left the boundary between the Soviet Union and the United States unresolved. Also, the regions of the Arctic could present problems in the future, especially between Canada and the Soviet Union in the outer Beaufort Sea, between Norway north of Spitsbergen, and between Denmark and the Soviet Union north of Greenland.

Chapter 2 details some of the problems currently experienced in the Soviet petroleum industry and includes a discussion on the recent downward trend in petroleum production and the reserves-to-production ratio, and the role of petroleum in the Soviet Union's export earnings.

In Chapter 3, a geological assessment of the offshore region is reviewed and includes a discussion and assessment of the reserve potential of the basins.

In Chapter 4, a detailed discussion of present activities, both administrative and technical, in the Soviet offshore waters is presented. From the point of view of actual activity in the offshore waters of the Soviet Union, this is the most relevant section. Considerable detail is afforded the technological requirements and purchases of offshore equipment. Dealings with foreign firms and nations in the development of the industry are also considered.

Finally, in Chapter 5, there is a return to the questions posed at the outset concerning the practical ability of the Soviet Union to undertake petroleum exploration and production in the offshore waters. The factors ultimately affecting the decision of Soviet planners are discussed, and an attempt is made to predict the way in which the offshore petroleum industry will develop and to make the reader aware of some of the industry investment issues.

What follows is a speculative yet detailed study on a sector of the Soviet energy industry which, to date, remains virtually unnoticed in the West, especially in the United States. The Soviet Union is developing by all the means at its disposal an offshore industry to rival one day that of the United States. If the assistance of Western powers, such as the United Kingdom and Norway, continues, then the issue of limited, present-day Soviet technology will gradually be overcome.

In addition, because the Soviet Union is concentrating its petroleum exploration efforts in the waters of the Arctic, the strategic and military consequence of Soviet experience and technical expertise in this region will surely lead to yet another zone of conflict between the United States, its allies and the Soviet Union. That dispute will be both sensitive and extremely interesting.

Notes

1. It might be of some interest to the reader that the total proven reserves in the Soviet Union are around 79 billion barrels. The range I have estimated is, therefore, quite speculative. The estimate of 98 to 151 billion barrels of potential reserves does not, however, come close to the estimate for proven reserves in the Middle East which was 401 billion barrels as of the end of 1987. Nonetheless, if the estimate I have quoted is true, it would still be substantial by world standards. See American Petroleum Institute, The Basic Data Book (Washington D.C., 1987), vol. 7, no. 3, Table 1, sec. 2.
2. "Soviet Union: Offshore Prospects and Problems," Petroleum Economist (May 1977), p. 191.
3. A.A. Meyerhoff, The USSR Northern and Far Eastern Coasts: Petroleum Geology and Technology, Mining Activities, and Environmental Factors (Unpublished paper prepared for the Department of Indian and Northern Affairs, Ottawa, Canada, 1980), p. 167.

2

Problems in the Soviet Petroleum Industry: Background to the Move Offshore

It is first necessary to put the discussion of offshore petroleum within the context of the operation of the whole Soviet petroleum industry. An enumeration of the problems that currently beset the Soviet petroleum industry will be given in this chapter. The intent is to make the reader aware of the salient issues facing Soviet planners in the oil industry and to discuss offshore petroleum within the context of energy planning in the Soviet Union.

The Slowdown in Soviet Petroleum Production

When the Central Intelligence Agency (CIA) released its April 1977 report on Soviet oil production, [1] the agency unleashed widespread doubts about the Soviet Union's ability to increase its oil output. The CIA predicted that, at worst, Soviet oil production would peak at around 11 million barrels a day (mbd) in 1978 and steadily decline thereafter. At best, output might not peak until 1980-1981 at around 12mbd and then fall off around 1982. Further, the agency claimed the Soviet Union would find it difficult to meet domestic and Eastern Europe oil demand while exporting to non-Communist countries at the same pace as in 1977. According to the CIA's worst-case scenario with output peaking at 11 mbd by 1978, the Soviet Union would be a net importer of oil by 1985.[2]

Even though the agency backed down from its forecast that the Soviet Union would become a net importer of oil by 1985, one should not reject the reasoning behind the agency's conclusions.[3] Indeed, the CIA was not that far off in its optimistic estimates of the Soviet Union's reaching only 12 mbd by 1980. Edward Hewett maintains this estimate was far better than the conventional wisdom of the time. Most other observers were claiming the Soviet Union would reach its planned 1980 target of 12.8 mbd.[4] The actual production was well below plan in 1980, reaching only 12.07 mbd.[5]

While some of the forecasts by the CIA may have been overly pessimistic, its reasoning was quite sound. Essentially the agency based its predictions on three features of the Soviet petroleum industry with which most experts agree:

i) *Overproduction at well sites through the early use of secondary recovery techniques (such as water injection) has resulted in less oil being ultimately recovered.*

ii) *Soviet emphasis on development drilling over exploration resulted in new discoveries that failed to keep pace with increasing output.*

iii) *If existing supply and demand trends for oil in 1977 were extended, Soviet oil production, given its 1977 rate of growth, would be unable to fulfill those requirements.*[6]

It is true that oil production in fact did not peak at 12mbd, but actually reached 12.33 mbd (616 mty) in 1983; the fact that output declined in 1984 to 12.23 mbd (614 mty) lends credibility to the CIA's pessimism.

Table 2.1 shows the Soviet Union's oil production from 1960 to 1987. What is especially important is the evidence of declining growth rates. Over the 1960 to 1970 period the Soviet Union sustained an average annual growth of 9%. Between 1970 and 1975 this had fallen to 7%.

Table 2.1
USSR Petroleum Production (1960-1987)

Year	Millions of tons	% Change	Year	Millions of tons	% Change
1960	147.9		1973	429.0	7.1
1961	166.1	12.3	1974	458.9	7.0
1962	186.2	12.1	1975	490.8	7.0
1963	206.1	10.7	1976	519.7	5.9
1964	223.6	8.5	1977	545.8	5.0
1965	242.9	8.6	1978	571.5	4.7
1966	265.1	9.1	1979	585.6	2.5
1967	288.1	8.7	1980	603.2	3.0
1968	309.1	7.3	1981	608.8	0.9
1969	328.3	6.2	1982	612.6	0.6
1970	353.0	7.5	1983	616.3	0.6
1971	377.1	6.8	1984	612.7	-0.6
1972	400.4	6.2	1985	595.8	-2.8
			1986	610.8	2.5
			1987	624.0	2.2

Average Annual Growth Rates in Percentages

1960-1970	9.08	1980-1982	0.78
1970-1975	6.81	1982-1984	0.01
1975-1980	4.20	1983-1985	-1.68
		1980-1985	-0.25

Sources: *Narodnoe khoziaistvo SSSR* (various years); 1987 estimate from *Tsentral'noe statisticheskoe upravlenie (TsSU)* cited in *Ekonomicheskaia gazeta* 1988 no. 5.

By the 1975 to 1980 period the average annual growth rate had declined to 4.2% and then fell even further between 1980 and 1982 to 0.78%. The negative growth from 1983 to 1985 of -1.68% evidenced real problems with Soviet oil production. Production in the first half of 1986 averaged about 12.2 mbd against 11.9 mbd for the first half of 1985, a sign of slight improvement.[7] Nonetheless, this output performance has meant that annual oil production goals have not been met for a decade. The increase in 1986 was still nearly 155,000 barrels a day (bd) below plan.

On the production side the real problem with the Soviet petroleum industry went beyond overproduction at well sites as the CIA pointed out. In addition, there was a problem of underexploration despite repeated warnings from officials in the Tiumen' province that underexploration would mean that Samotlor would be responsible for ever-increasing shares of the overall petroleum and energy demand. Also, failure to upgrade production techniques to extract oil as it became less accessible through over-exploitation meant that planners were not even facilitating the short-term goals of increased output.

As oil production moved farther east from the Volga-Ural region in the 1960s, production costs inevitably rose. The fields of West Siberia's Tiumen' oil province at present account for around 60% of total oil production. These fields--notably Samotlor and the Tiumen's other two big fields, Fedorovo and Mamontovo--are nearly at their upper limit. The only choice for the Soviet Union to increase output is to move into the more remote and more costly fields farther east. Many Soviet analysts maintain that the petroleum industry has become more labor intensive, and the inability to attract workers to areas that have extremely long winters and short summers, during which much of the region turns to swamp, has meant a lot of problems for the industry.[8]

The new fields brought into production in the 1970s were located in the Middle Ob basin. However, probably for the reasons given above, the burden of meeting output targets was placed on the Tiumen' district, especially Samotlor. The problem was that Gosplan, the Soviet State Planning Commission, did not step up exploratory drilling; instead, it continued to put more faith in Samotlor's ability to deliver increased oil production. Thane Gustafson points out that in

1975 most Tiumen' oilmen had reached the opinion that Samotlor would peak around 1977 at 100 mty. Only after considerable pressure from above was brought to bear upon them did they reluctantly increase their 1980 estimate to 120 mty.[9]

The result was a shortfall of the plan for geological exploration, as was predicted by Farman Salmanov, chief of the Tiumen's geological service. He writes, "...to turn prospects into reality would require bringing annual exploratory drilling in West Siberia up to two million meters a year, a fourfold increase."[10] As a result, West Siberian geologists failed to meet their plan targets in 1976, 1977 and 1978.

Planners were trying to extract more oil from the Tiumen' district largely because of drastic shortfalls in production in other areas of the country. Additions of new technology were not forthcoming until the early to mid-1980s. One *Pravda* article, written as late as September 1985, pointed out that the country's industrial enterprises were not turning out new equipment for "...accelerating the sinking of wells and derricks in complete sets with the necessary auxiliary equipment, which would raise labor productivity. "At present," the article reveals, "the oil workers have to make this equipment themselves."[11]

The new capacity that was being eked out of Samotlor was not offsetting the decline in other regions. Failure to undertake serious exploratory drilling in the western and eastern Siberian fields and failure to upgrade extraction technology are among the primary reasons for the declining output in 1984. These failures explain the current dilemma in which Soviet energy planners find themselves.

Continued Emphasis by Planners on Oil Production

As the problem became more evident to Soviet planners in the late 1970s, the logical response one expected would have been a shift of emphasis and investments to other energy sources and an increase in energy conservation, much the way western Europe dealt with its problems after the OPEC oil shock. Instead, as shown in Table 2.1, Soviet oil production pushed ahead through to 1984, when it experienced its first decline and then picked up again over

1985 and 1987. The appearance of stability is, however, deceptive. As Gustafson points out, the only reason Soviet oil output remains at its current levels is due to the Soviet leadership's "...willingness to pay a steadily increasing price for it, and the price is growing at a breathtaking pace."[12]

Based on cost estimates for 1982 from *Narodnoe khoziaistvo*, it was found that for the 8.7 billion rubles of capital invested in oil extraction, 3.78 million tons of oil equivalent (toe) was yielded. This compared with 6 billion rubles invested in gas which yielded 29.4 million toe and 2.2 billion rubles invested in coal which yielded 6.3 million toe. In other words, the resources required to produce oil are almost twice as high as those for gas and coal.[13]

When Yuri Andropov came to power in 1983, there were signs that something concrete was to be done about the problem. *Pravda*, in April of that year, announced that a new energy plan was to be drawn up and released. The report was not released that year, but from statements made by Andropov and from a report by the chairman of the Academy of Sciences of the USSR, A.P. Aleksandrov, the emphasis was to have been on energy conservation and oil displacement. Months passed, however, and no report emerged. Some observers believe that because Andropov was ailing, the head of the oil industry, V.I. Dolgikh, reemerged. Dolgikh had been one of the strongest to advocate that abundant present supplies of oil were more important than conservation. The energy plan was published only two months after Konstantin Chernenko came to power. Instead of concentrating on the demand side of oil, it emphasized the supply side, by calling for continuous growth in output through to the year 2000.[14]

When Gorbachev replaced Chernenko, some expected that he would follow Andropov's path of trying to do something about the demand side of petroleum. Instead the contrary occurred, at least in the initial months of his accession to power. In a speech given to the workers of the Tiumen' oil district, Gorbachev noted:

> *One must always keep in mind that the success of the workers of Tiumen' strengthens the economy while their failures cause considerable slowdowns. For this reason, the CPSU Central*

> *Committee is bothered by the fact that the Tiumen' region has not fulfilled its plan for the last three years. This failure creates difficulties in the economy.*[15]

Gorbachev was essentially echoing the concerns of the Brezhnev era. To be fair, at a conference in Tiumen' opened by Gorbachev, some of the discussion centered around the technical needs of oil extraction, but little emphasis was placed on development drilling.[16]

It would appear these were not simply words to appease the old guard who may have jealously held onto old beliefs of how the oil industry should run. Several articles appeared in *Ekonomicheskaia gazeta* after the 27th Party Congress in February, 1986 on the need to increase output in West Siberia in order to meet the production goals set down at the congress. In one article the problem was stated as follows:

> *In 1985, Nizhnevartovskneftegaz unsatisfactorily utilized the fund for oil wells. There was important neglect of the Samotlor deposits and output of technology causing serious shortcomings in the organization of work. Now the affairs at Samotlor are corrected--the most important oil field of the Tiumen' district has found its second wind.*[17]

Soviet planners are focusing, therefore, on the supply side rather than on the demand side. This was made evident when N. Ryzhkov, chairman of the USSR Council of Ministers (or prime minister), announced at the party congress that capital investment in the oil and gas industry will be stepped up by 47% during the twelfth five-year plan (FYP) to ensure reliable supplies of fuel.[18]

The 27th Party Congress and Petroleum Production

There are, nonetheless, signs that the new administration is not adhering strictly to traditional approaches. While officials are doing little to attack the demand side of oil, they are realizing the need to undertake more exploratory drilling, especially in West Siberia. This represents an improvement over the position expressed by Gorbachev on his visit to the Tiumen' district in the fall of 1985. The Ministry of Oil, which undertakes only one-third of the oil exploration and production in the Soviet Union (the remainder of which is done by the Ministry of Geology and the Ministry of Gas), has a 1986-1990 plan target for drilling 60% more than their 1981 to 1985 level, of over 200 million meters.[19]

Therefore, the emphasis is still on petroleum production, albeit in a more systematic fashion, with increased concern now for the technological level of extraction and for development drilling. Clearly, this has had some effect as can be seen from the high output figure for 1987 cited in Table 2.1. This is, however, due to stricter discipline at field sites (as Gorbachev's Tiumen' speech suggested would happen). The increase is not, however, a solution and may do more to exacerbate the problem than help. It would seem more oil is being produced through extensive development rather than the necessary intensive development.

The targets set at the 27th Party Congress are extremely high when one considers that other energy sources are almost twice as cheap to produce. As can be seen from Table 2.2, the party congress called for between 625 and 640 mty oil (more than has ever been produced), 835 to 850 billion cubic meters (bcm) gas and between 780 and 800 mty coal.

Table 2.2
Forecast of Oil, Gas and Coal Production According to Projections in the Twelfth Five-Year Plan, 1986-1990

Oil:	625 - 640 mty
Gas:	835 - 850 mcm
Coal:	780 - 800 mty

Source: *Ekonomicheskaia gazeta* (March 1986, no. 12).

In the coming years it appears the Soviet Union is preparing for a higher growth in petroleum production than experienced in the 1980 to 1984 period. Table 2.3 shows the average projected rate of growth in fuel energy from 1980 to 1990. If the high plan targets between 1985 to 1990 are to be met in petroleum production, a 0.73% growth rate over this period, as compared to a 0.39 growth rate over the 1980 to 1984 period, is required.

Table 2.3
Average Annual Rate of Growth of Fuel Energy, 1980 - 1990 (selected periods)

	Oil	Gas	Coal
1980-1984	0.39	7.78	-0.14
1983-1984	-0.58	9.65	-0.53
1984-1990* (low)	0.33	6.03	1.52
1984-1990* (high)	0.73	6.35	1.95

* The rate of growth required to attain the planned output as proposed in the Twelfth Five-Year Plan.

Sources: *Narodnoe khoziaistvo SSSR* (1986), *Ekonomicheskaia gazeta* (March 1986, no. 12).

Even more revealing is that the requirements for growth in gas output are lower over the next five years than over the 1980 to 1984 period. Again, taking the high estimate of gas production to 1990, gas output would only have to grow at a rate of 6.35% compared to a growth of 7.78% over the 1980 to 1984 period. All indications are then that the Gorbachev administration is pursuing a familiar policy of increased oil production. The question, therefore, is why? Indicators suggest that the marginal cost of extraction is climbing at an alarming rate and returns on investment are decreasing. Further, by not attacking the demand side, planners are sure to face massive waste of fuel as in the past. Gas is more abundant than any other fuel, yet it appears that the emphasis is on increasing oil output growth rather than gas.

Gorbachev is surely not naive about the problems outlined above and, given the choice, he would probably like nothing more than to back down from the output requirements the twelfth FYP has called for; but his hands, like Brezhnev before him, are tied. The Soviet Union must continue to produce oil if the economy is to function at a reasonable pace over the next plan period.

While gas production has shown steady increases over the past decade, gas is still not the major source of fuel for either the industrial or commercial sectors of the economy. Because new fields are more remote, the problem of transporting gas to consuming areas presents more of a problem than transporting oil (due to pipeline requirements and compressor apparatus). Coal, also a traditional energy supplier, is in more trouble than oil. Coal production has decreased steadily over the past decade even though the poorer grade of coal now being mined in areas east of the Donets basin demands that there be more production, not less, to meet the same energy needs. Coal production swelled to 718 mty in 1982, but declined after that to 712 mty in 1984.[20] If coal production hovers around the 700 mty mark, the lower caloric content of this fuel can only increase the need for more petroleum.

The Soviet Union has no other hard currency earner that compares to oil. While the problems of energy production are important, nowhere is it more seriously felt than in the balance of payments. The inability to produce manufactured goods for export leaves the planners under Gorbachev with no other choice but to increase petroleum production.

The Importance of Soviet Petroleum Exports

One of the main criticisms of the CIA 1977 report is that its prediction of the Soviet Union becoming a net importer of oil has not been realized. The problem with the CIA findings is that they ignore the fact that the Soviet Union cannot afford to become an oil importer. One telling calculation notes that if the Soviet Union were to import 1.0 mbd oil in 1990, given a constant relationship between oil and gas world market prices (WMP), it would cost the

Soviet Union its entire revenue of 60 bcm natural gas.[21] The loss of revenue by not exporting petroleum would cut hard currency earnings by over 60%, leading to an enormous reduction in non-oil imports.

The Soviet Union has become dependent upon petroleum exports in large part because of its inability to replace oil with another hard currency earner. Soviet finished goods are in little demand beyond the Council of Mutual Economic Assistance (CMEA), and raw materials, such as gold and other metals--owing to the lack of cartels--are subject to huge price swings, which makes them unpredictable hard currency earners. Petroleum, through the surges and recessions of the Western economies, has served the Soviet Union as a balancing mechanism preventing the trade deficit from becoming too serious. Marshall Goldman points out that petroleum is used for this purpose even when so doing occasionally produces a serious effect on the domestic economy.[22] What is curious about the Soviet export issue is the one-sidedness of the problem. In absolute terms, the extent to which West Europe and the rest of the Organization for Economic Cooperation and Development (OECD) countries rely on Soviet oil imports is marginal. The Soviet share of Western European oil imports has never exceeded 7% of the region's total oil imports.[23] In terms of total OECD oil imports, Soviet oil constituted only 6% in 1984 and is thought to be a still smaller share in 1985 and 1986.[24] On the other hand, as a number of authors have pointed out, the extent to which the Soviet Union relies on those same regions is enormous. In 1979, almost two-thirds of Soviet exports to the West were in energy-related products and over 50% of these exports were related to oil. By 1984, as illustrated in Table 2.4, petroleum and petroleum products' share of the value of total Soviet trade with the OECD was 62%. The problem seems to have been further aggravated in 1985 with the total value of Soviet hard currency earnings from petroleum sales to the OECD declining while the share of petroleum exports in total export earnings increased to 63%.

The importance, therefore, of Soviet petroleum exports to the Soviet economy is then expressed in these figures. However, it is necessary to see the full implications of such a dilemma and to trace the historical development of the problem if it is to be fully understood.

Soviet Energy Export Earnings: Petroleum

In terms of the value of Soviet exports to hard currency countries, Table 2.4 illustrates the problems the Soviet Union has encountered in recent years. With the fall in price per barrel of oil from an average of $36.25 in 1981 to $29.00 in 1984 and then to around $15 in 1986, it was necessary for the Soviet Union to increase substantially the quantity of oil sold in order to maintain a steady growth in export earnings. This is important when considering that between 1978 and 1981 the actual quantity of petroleum exported fell, while the value of gross oil exports steadily rose from $6.3 billion (1978) to $13.9 billion (1981). Prior to 1981 the Soviet Union incurred high earnings from petroleum at a relatively low volume of exports. After 1981, however, it became necessary to increase markedly the amount of petroleum exports, simply to maintain a steady increase in hard currency earnings. This spells out the problems for the Soviet Union.

Between 1981 and 1984 actual quantities of oil exports rose from 52.57 mty to 81.4 mty. Assuming that, at a minimum, Soviet planners wished to earn in 1985 and 1986 the $16.9 billion earned in 1984, it would have been necessary to sell 85.1 mty in 1985 (with oil at its 1985 average price of $27 per barrel) and 155.1 mty in 1986 (at the 1986 average price of $15 per barrel). Clearly, this was not the case. In 1985 it is estimated that Soviet oil exports were around 55.1 mty, earning approximately $13.4 billion, or a loss of $2.8 billion over 1984.

This fall in revenues is reflected in official Soviet statistics, where total value of exports to the "developed capitalist countries" fell from 21.3 billion rubles in 1984 to 18.6 billion rubles in 1985, a decline of 2.7 billion rubles.[25] A large part of this official decline in export earnings is explained by the decline in oil exports to the OECD which is reported by Soviet sources to have declined by 2.21 billion rubles in 1985.[26]

If the Soviet Union exported only as much as 55.1 mty in 1986 (which it did not), the loss incurred due to the fall in oil prices to $15 per barrel in the last year would mean that the Soviet Union could have earned less than $6.0 billion dollars in petroleum export earnings in 1986. It can be argued that since oil prices are settled in US dollars, and that the dollar remained fairly strong until August 1986, the Soviet Union had a windfall gain because of the inflated dollar. This, however, is hardly comfort to the Soviet planners, who in both nominal and real terms witnessed a drastic slide in export earnings. When the American dollar collapsed in the last quarter of 1986, Soviet export earnings were further reduced.

Table 2.4
Value of Soviet Petroleum and Petroleum Product Exports to the OECD and Total Value of Soviet Exports to the OECD, 1971-1986

(Millions of US $)

Year	Petroleum/ Petroleum Products	Total Exports to OECD	% of Oil Exports to Total Exports	Price per Barrel
1971	846	3,060	28	-
1972	880	3,161	25	-
1973	1,421	4,889	27	-
1974	3,105	7,920	37	-
1975	3,782	8,494	43	-
1976	5,173	10,765	48	-
1977	5,777	12,128	47	-
1978	6,380	13,095	46	-
1979	10,110	18,568	50	-
1980	13,449	23,789	54	32.8
1981	13,903	23,767	57	36.2
1982	15,549	25,060	61	33.9
1983	15,891	24,749	64	29.9
1984	16,140	25,893	62	29.0
1985	13,368	21,143	63	28.0
1986	-	-	-	15.0

Source: OECD, *Foreign Trade By Commodities*, (Paris 1987) vol. 2, sec. 33.

Observing the value of petroleum exports sold to the OECD over the 1971 to 1985 period in Table 2.4 and comparing them as a percentage of total Soviet trade with the OECD, the results are alarming. Between 1971 and 1980 petroleum as a share of total exports rose from 28% to 54%. This may in a small way reflect the increasing inability of Soviet manufactured goods to compete on world markets, a situation with which Soviet planners are not comfortable. The planners, however, could at least take consolation in the increase in oil production over this period, thus alleviating some concern over hard currency sources. But from 1981 to 1983, Soviet planners had no such consolation at hand. As a share of total exports to the OECD, petroleum rose from 57% in 1981 to 64% in 1983, and remained near that level at 63% in 1985. Comparing this to the trend displayed in Table 2.1--where petroleum production fell between 1983 and 1984 by -0.6% and increased in 1986 and 1987 by an average of only 2.35% contrasted with the average yearly increase of 4.7% over the last half of the 1970s when their production problems had set in--it would appear that petroleum as an hard currency earner is threatened. The point to make here is that over the 1970s and the early 1980s the Soviet Union's rapid rise in oil export earnings was primarily due to the steady rise in oil prices. It was in this time that the real dependency upon petroleum as a hard currency earner became evident. As a result, when petroleum production began to fall off in 1983, the Soviet Union was not in a position to replace that hard currency earner with any other commodity. It was necessary, therefore, to continue increasing the export of petroleum even in the face of declining production. As they have increased production in the years 1986-1987 at a pace I do believe sustainable, the price per barrel of oil has not returned to anything near the $30 a barrel mark that gave the Soviet Union its huge hard currency earnings of the late 1970s and half of the 1980s. To date this hard currency source has not been replaced.

Soviet Energy Export Earnings: Gas

One of the overriding concerns of Soviet foreign trade planners is the role of natural gas as an export earner and its ability to someday replace oil in this crucial role. Since 1971, natural gas has risen rapidly as a share of total exports to the OECD, rising from a share of total exports of less than 1.0% in 1971 to over 13% in 1983. This, no doubt is an indication of the priority given to the gas industry throughout the 1970s.

Table 2.5
Exports of Natural Gas to Major OECD Recipients 1980-1985 (mcm/day), prices/toe, US dollars; Consumption of Natural Gas in EEC, 1980-1985 (mcm/day)

	1980	1981	1982	1983	1984	1985
Austria	5.7	8.5	8.5	11.9	11.0	11.5
Finland	2.8	2.8	2.8	1.9	2.2	2.7
France	11.3	17.0	11.3	11.0	12.4	18.6
Italy	19.8	22.7	25.5	23.3	20.5	16.4
FRG	31.2	34.0	28.3	30.4	34.0	34.0
Total	70.8	85.0	76.2	78.4	80.0	83.2
Price/toe						
France	108.2	127.4	127.3	131.3	127.5	125.2
FRG	103.0	128.7	141.0	128.9	123.0	113.4
Italy	100.1	138.9	152.2	141.4	147.4	-
EEC Consumption						
	546.5	536.7	513.6	542.7	572.0	-

Sources: Natural Gas Exports (1980-1983) from CIA, March 29, 1987. (1984-1985); from Jonathan P. Stern, "Soviet Oil and Gas Production and Exports to the West: A Framework for Analysis and Forecasting," Joint Economic Committee, *Gorbachev's Economic Plans*, vol. 1 (Nov. 23, 1987), pp. 500-513. Price/toe from IEA, *Energy Prices and Taxes*, (Q1) 1985. EEC consumption from OECD, *World Oil and Gas Statistics*, (Paris, 1987).

What is interesting about the results in Table 2.5, is the decline of gas as a share of total trade and value of export earnings since 1981. As a share of total exports, gas fell from 16.8% in 1981 to 13.3% in 1983. The corresponding value in export earnings fell from $4.1 billion in 1981 to $3.3 billion in 1983. The reason for this decline can, to a large extent, be explained by a decrease in West European demand for imported gas and the subsequent fall in price. In this period, earnings from gas fell at an average annual rate of -10.3%.

Soviet gas exports to West Europe in the last five years show a downward trend. Combined, Finland, France, and Italy are the West European countries to which the Soviet Union exports most of its gas. In each of these three cases, the quantity of gas exported fell between 1982 and 1983. While Finland and France increased slightly their imports of natural gas in 1984 and 1985, Italy--traditionally a much bigger buyer of gas than either Finland or France--has steadily decreased its gas imports from the Soviet Union to around 16 million cubic meters (mcm) a day in 1985. In the case of the West Germany (FRG), the largest single buyer of Soviet gas, the increase to importing 34 mcm per day in 1985 (a return to their 1981 level) is due in large part to the rapid decline in price per ton of oil equivalent (toe) between 1981 and 1985. This translates into very little change in export earnings from gas.

The portion of Soviet gas to total European Economic Community (EEC) consumption has remained fairly steady at 13% in 1980 and up to 14% in 1983 (a slight decrease from the 1981 level of 16%). When this is considered in conjunction with the price of gas, which--except for Italy--fell in 1985 to an average of $119.32 per toe, its lowest level since 1980, then Soviet planners must be concerned.

Given this assessment, gas cannot be relied upon to replace oil as an export earner. That is, it cannot play a role equivalent to oil in the Soviet hard currency balance of payments. It was pointed out by McMillan and Hannigan as early as 1983, that gas exports will "...remain substantially tied to purchases on credit of material and equipment for the development of the production and distribution capacity of the Soviet gas sector."[27] Since more gas exports rely on more gas production which in turn relies upon purchases of Western equipment, then the extent to which gas exports represent new income is difficult to gauge.

While it can be argued that gas is being used to satisfy domestic and CMEA demand, there are signs of strain even in this regard. As shown in Table 2.6, the average annual growth in gas production between 1970 and 1975 was 7.9%. This rate increased between 1975 and 1980 to 8.5%, representing the substantial priority granted to the gas sector over this period. In the last five years, however, complications began to appear in the production statistics. Between 1980 and 1984 gas production, as reported by official Soviet sources, slowed to 7.8%. Between 1984 and 1986 there was only a slight average annual increase of 8.1%. Soviet officials claim they have discovered enough gas to last well into the 1990s.[28] The issue at present, however, is the increasing cost of production in the gas industry. With most of the new gas discoveries being north of 60 degrees latitude and farther and farther east, pipeline and compressor costs are very dear.

Table 2.6
Soviet Gas Production, 1940 - 1987
(mcm/year)

1940	3.2	1982	500.7
1960	45.3	1983	535.7
1970	197.9	1984	587.4
1975	289.3	1985	643.4
1980	435.2	1986	686.0
1981	465.3	1987	727.0 (plan)

Average Annual Growth Rates
(%)

1970-1975	7.9
1975-1980	8.5
1980-1984	7.8
1984-1986	8.1
1980-1986	8.1

Source: *Narodnoe khoziaistvo SSSR*. 1987 plan target from *TsSU* cited in *Ekonomicheskaia gazeta*, 1988 no. 5.

One of the most revealing indications that gas will not be used as a replacement for oil as a hard currency earner is the increase in gas exported to East Europe. A March 1987 report issued by the CIA claims that Soviet gas exports increased from 84.1 mcm per day in 1981 to 97.6 mcm per day in 1983. In all likelihood increased exports of gas to the CMEA were an effort to release oil for hard currency exports. This leads to an important point when considering the importance of Soviet petroleum exports: notably, the substantial quantities of oil exported by the Soviet Union to the CMEA.

Soviet Petroleum Exports to the CMEA

Between 1980 and 1986, the Soviet Union exported to the CMEA an annual average of 90 mty of oil. This is an increase over the 1975 to 1980 average of 79 mty. While oil exports have decreased since their high in 1981 of 95.5 mty, it is clear there is a minimum acceptable level to which oil shipments to the CMEA can sink. Between 1981 and 1984, according to the eleventh FYP, oil shipments were not to exceed 80 mty, but in fact averaged 90 mty as shown in Table 2.7.

This has two important implications for the Soviet petroleum industry. First, while the Soviet Union may be trying to increase gas exports to the CMEA to free-up oil for export, it can only be done slowly and to a limited degree. Second, because there is a minimum level of oil exported to the region, Eastern Europe will always be a factor for the Soviet Union to consider when it undertakes petroleum production, both from the standpoint of a drain on hard currency earnings and as a constituent requiring a permanent share of the Soviet Union's total petroleum production.

The Soviet Union, as may be apparent, is in a dilemma. As oil production has declined over the past three years, its importance as an earner of hard currency has not diminished. Gas production, while significant, is unlikely to replace petroleum in supplying domestic energy demands and certainly cannot replace it as an export earner. Although East Europe is warned constantly oil shipments to it are about to be cut back, around 15% of Soviet oil production is always allocated to East Europe nontheless.

Table 2.7
Soviet Exports of Petroleum to the CMEA (including Cuba, Vietnam and Mongolia), 1975 - 1986

	mty	% change
1975	73.1	-
1976	79.4	8.6
1977	82.1	3.4
1978	86.6	5.5
1979	91.1	5.2
1980	92.6	1.6
1981	95.5	3.1
1982	87.4	-8.5
1983	87.4	0.0
1984	88.9	1.7
1985	85.4	-3.9
1986	91.8	7.5

Source: CIA, *International Energy Statistical Review* (March 29, 1987), p. 16.

The fields of West Siberia are responsible for an ever-increasing share of total petroleum production. This compensates for the decline in other areas of the country. The problem, however, is that there have been no significant additions to the reserves since 1973. As the reserves-to-production ratio falls, the majority of oil production is seen to be coming from fields approaching exhaustion.

The Soviet Policy Response

A number of Soviet economists have warned for the last six years that further increases in oil production would be impossible over the long run if West Siberian depletion ratios (the gross increase in Siberia's output required to compensate for declines elsewhere in the region) climbed much above their 1980 levels of 49%. One Soviet estimate (S.N. Staroitov) has put the 1985 depletion ratio in West Siberia at 86%.[29] This decline was predicted by Academician Trofimuk. He commented as early as 1981 that the decline in the reserves-to-production ratio in West Siberia "...has been alarming and indeed more rapid than anywhere else in the country."[30]

Table 2.8
USSR Reserves-to-Production Ratios for Selected Years Based on Varying Estimates (proved and probable)

Year	Soviet	CIA	UN	International Petroleum Encyclopedia
1970	57/1	29/1	23/1	32/1
1975	41/1	20/1	17/1	23/1
1977	37/1	18/1	15/1	21/1
1979	34/1	17/1	14/1	19/1
1980	33/1	17/1	13/1	19/1
1984	33/1	16/1	13/1	18/1

Notes: The reserve estimates were chosen to give a broad range from often cited sources.

Sources: Marshall Goldman, *The Enigma of Soviet Petroleum* (1980). Estimates for 1984 based on production in Table 2.1.

Table 2.8 shows reserves-to-production ratio estimates from 1970 to 1984. While the estimates vary, the decline since 1970 is an annual average of 3.9%. This decline must have been deemed unacceptable by planners as well as academics. Although exploration was cut back in the first half of the 1970s, discovery rates began to pick up again in the last half and a total of 41 new fuel deposits were found.[31]

The amount of exploration drilling necessary, however, has not been enough when compared to the commitment given to the development and production of petroleum. Exploration as a share of total spending on oil and gas declined from 32% in 1980 to 20% in 1982. This was seen as insufficient to support the 19% increase in exploratory drilling called for in the eleventh FYP.[32]

There are indications that many geologists are becoming very pessimistic about the ability of West Siberia generally and the Tiumen' district specifically to add to reserves. Northern Tiumen', long thought to have huge oil reserves, is suspected of being mostly condensate [33] and in the eleventh FYP, northern Tiumen' was ranked last in priority by the Ministry of Geology for exploration. Increasingly, West Siberia is viewed as being near its limit. A.A. Makarov suggests that output from West Siberia will drop precipitously after 1990 if attempts are made, as in the past, to boost oil production.[34]

The message seems to have sunk in, judging by statements made by Gorbachev and Ryzhkov in 1985 and 1986. Gorbachev's first policy statement on petroleum came in the fall of 1985 when he toured the Tiumen' district and, among other things, criticized the branch science research organizations of West Siberia and the Ministry for the Construction of Petroleum and Gas Industry Enterprises. He called upon them to improve their extraction techniques and to provide better forecasts for the industry, as well as prepare for an expansion in the volume of construction and installation work.[35]

The push initiated by Gorbachev seems to have had its effect on the production, but not the exploration of reserves. Tiumen' in 1986 made impressive gains and put a record number of fields on stream, increasing production to over 7.5 mbd.[36] While this was nearly 154,000 bd short of the target, it may have afforded planners the leeway needed to make concrete plans for exploratory drilling. In the past, due to either bad planning or necessity, oil production goals took precedence over exploration goals. But, since the 27th Party Congress in February of 1986, it appears this may be reversing. Ryzhkov has announced that exploratory drilling is also an important part of the overall investment in the energy sector. According to Ryzhkov the development of the Soviet economy requires large-scale exploration for oil and gas and a higher level of efficiency and quality in preparing reserves for production.[37]

The Soviet leadership, it would seem, is trying to increase exploration but maintain the current level of production at the same time. Indeed, as has been illustrated, the leadership cannot afford to allow petroleum production to decline as it did from 1983 to 1986. Officials are, despite the risks, embarking on a twofold plan of increased production and exploration. What is interesting is that the pessimism expressed by academicians such as Trofimuk and Makarov about West Siberia seem to have been echoed in the twelfth FYP proposals for petroleum development. This statement in *Ekonomicheskaia gazeta*, reporting on the contents of the twelfth FYP, is revealing:

> *Oil production by the year 1990 will reach 625-640 mty (including gas condensate). Oil production will continue to grow in West Siberia, Kazakhstan and the north European parts of the country. Exploration of the deep oil deposits at the Prikaspiiskoe basin will begin. The opening up of oil and gas deposits on the continental shelf will be accelerated as will be the development of the necessary technology for this.*[38]

For the first time, the new FYP has called for the exploration and development of the continental shelf offshore. The reasons for this are quite simple. As the need for more petroleum is essential and the future of West Siberia looks dim in the view of many Soviet geologists, the potential of the remnant seas (Caspian, Black and Azov), the eastern waters of the Okhotsk and Bering seas, and the huge potential of the Soviet Arctic, including the Barents, Kara, Pechora and Chukotsk seas, is extremely attractive. This option is being undertaken as the best way to fulfill the twofold strategy of increasing production and exploration. Soviet planners can continue to increase output in West Siberia in the hopes that new reserves offshore will significantly increase their petroleum reserves-to-production ratio.

Notes

1. CIA. Prospects for Soviet Oil, National Foreign Assessment Review Agency ER-7710270 (Washington, D.C.), April, 1977.
2. CIA (April, 1977), p. 8.
3. Explanation of the CIA's revised estimates in CIA. The World Oil Market in the Years Ahead National Foreign Assessment Review Agency, ER-79-10327U (Washington D.C., Aug. 1979).
4. See: Edward Hewett, Energy Economics and Foreign Policy in the Soviet Union, the Brookings Institute (Washington D.C., 1984), p. 25. Hewett cites the conventional wisdom of the time, which is a useful source: George Hoffman, "Energy Projections - Oil, Natural Gas and Coal in the USSR and Eastern Europe," Energy Policy, vol. 7 (Sept. 1986).
5. Narodnoe khoziaistvo SSSR (1981).
6. CIA (April 1977), p. 1. See also: Marshall I. Goldman, The Enigma of Soviet Petroleum, (London: George Allan and Unwin, 1980), p. 114.
7. First half 1986 estimate from: Oil and Gas Journal (Aug. 4, 1986), p. 80.
8. Petroleum Economist (April 1985), p. 131.
9. Thane Gustafson, "Soviet Oil Policy and Energy Politics, 1970-1985" (mimeo), United States, 1986.
10. Farman Salmanov, cited in Thane Gustafson (mimeo), p. 41.
11. Pravda, Sept. 5, 1985, p. 1.
12. Thane Gustafson (mimeo), p. 11.
13. Thane Gustafson (mimeo), p. 9.
14. Thane Gustafson (mimeo), has an excellent discussion of this, and the reader is referred to his pp. 12-15 for a fuller explanation.
15. Gorbachev, Speech in Pravda, Sept. 7, 1985, pp. 1-2.
16. Conference of the Party and the Economic Aktiv of Tiumen and Tomsk Provinces. See Pravda, Sept. 7, 1985, p. 1.
17. Ekonomicheskaia gazeta, April 1986, no. 18, p. 1.
18. N. Ryshkov, quoted in Oil and Gas Journal, March 31, 1986, p. 8.
19. "USSR, No. 1 in Production, Could Claim Top Spot in Drilling," Oil and Gas Journal, August 25, 1986, p. 20.

20. Narodnoe khoziaistvo SSSR, 1984.
21. This calculation is based on the existing price relationship between oil and gas in 1980, see: Friedman Muller, "Implications for the West, including Defence, of the CMEA Energy Prospects," CMEA Energy 1980-1990, NATO Colloquium (April 8 - 10, 1981), pp. 303-320.
22. Marshall I. Goldman (London, 1980), p. 94.
23. John B. Hannigan and Carl H. McMillan, The Soviet-West European Relationship: Implications of the Shift From Oil to Gas, East-West Commercial Relations Series. no. 20, (Ottawa: Institute of Soviet Studies, Carleton University, 1983), p. 80.
24. "Oil Exports Rise Again," Petroleum Economist, February, 1985, p. 57.
25. Vneshniaia torgovlia SSSR, 1985, Table 3.
26. Vneshniaia torgovlia SSSR, 1986, Table 15.
27. Hannigan and McMillan, (1983), p. 82.
28. "Soviet Union: Survey of Energy Sources," Petroleum Economist, (April 1985), p. 131.
29. Thane Gustafson, (mimeo) p. 57.
30. Academician Trofimuk cited by Boris Rahmer, "Uncertain Prospects for Soviet Oil," Petroleum Economist, July 1981, p. 294.
31. David Wilson, Soviet Oil and Gas to 1990 EIU Report No. 90 (1980) p. 47.
32. Thane Gustafson, (mimeo) pp. 58-59.
33. Not that condensate is a particularly bad find. It requires far less refining than oil or high sulphur content gas and is becoming increasingly more valuable as its properties are discovered. But the point for Soviet energy planners at this time is to find petroleum.
34. A.A. Makarov cited in Thane Gustafson, (mimeo), p. 62.
35. M.S. Gorbachev, Pravda, Sept. 7, 1985, pp. 1-2.
36. "Soviet Rebound Boosts 1986 Communist Bloc Oil Output," Oil and Gas Journal, (Sept. 8 1986), p. 28.
37. N. Ryzhkov cited in Oil and Gas Journal, (March 31, 1986), p. 59.
38. Details of the 12th FYP, in Ekonomicheskaia gazeta, no. 12 (March 1986), p. 11.

3

The Geology of Soviet Offshore Petroleum: An Overview of Offshore Potential

This chapter will address the potential of Soviet offshore petroleum reserves for solving the problems that beset the petroleum industry. An enumeration of the geological potential of the Soviet shelf is intended to make the reader aware of the vast petroleum reserves of the Soviet shelf as well as some of the tenuousness with which these reserves are discussed. The discussion on climatic conditions is found in Chapter 4 as part of the analysis of technological requirements along with a section on administration of the offshore waters.

By generally knowing the resource potential of an area it is easier to predict where and how exploration and development will begin. The administration of the offshore petroleum industry, the technological requirements, and the political factors involved in deciding to go offshore are connected to this.

For example, going offshore in the northern region of the Bering Sea (in the Anadyr basin) presents technological difficulties of far greater proportions than going offshore in some other area such as the Caspian Sea or even the Barents Sea. This is due to the relative remoteness and excessive ice floe in these waters. Moreover, the Anadyr

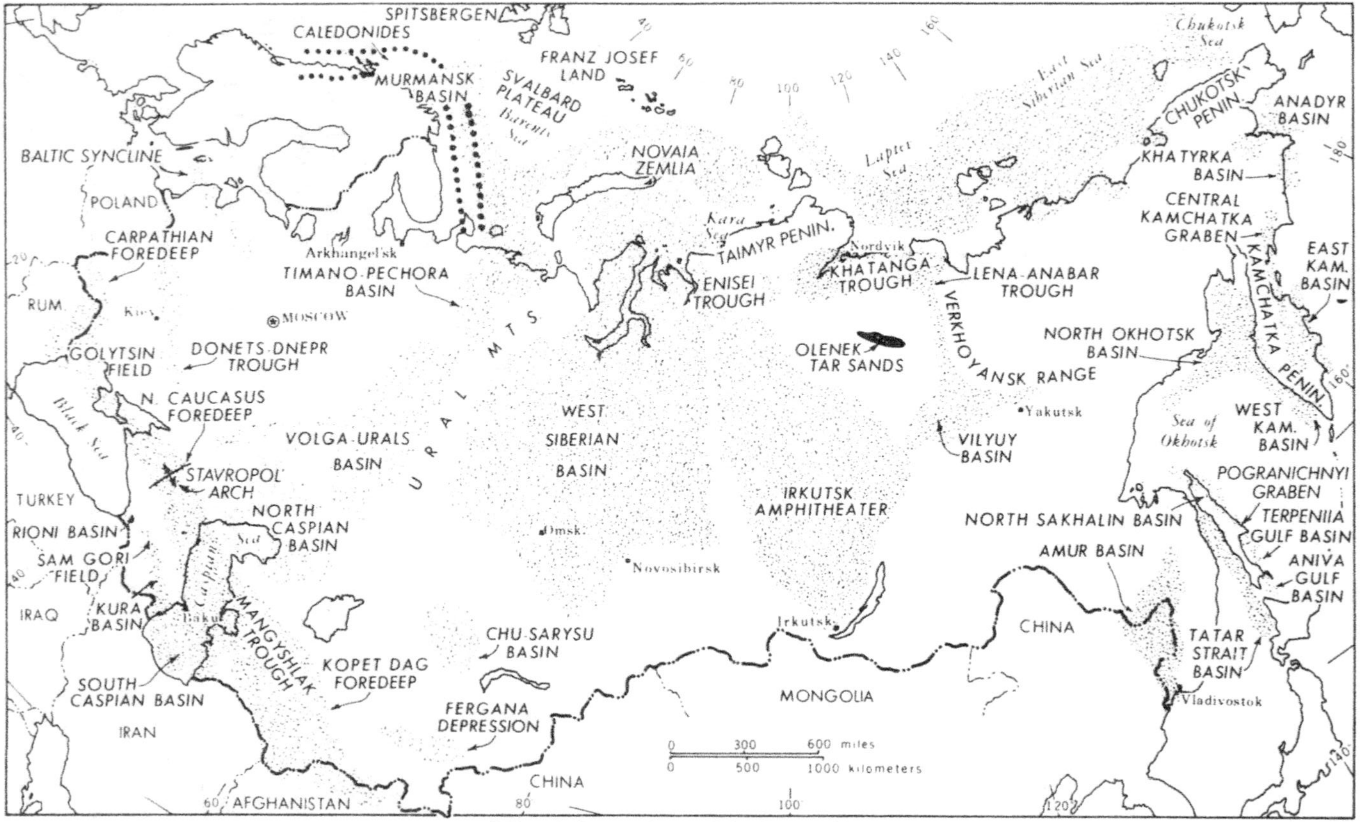

1. Oil and Gas regions of the Soviet Union

basin is predominantly a gas trap, and not even in commercial quantities. It is not necessary, therefore, for the Soviet Union to spend large amounts of hard currency on, or to develop for themselves at considerable expense, the technology required to explore the northeastern region. Further, by going offshore in the Caspian Sea or the Barents Sea, the Soviet Union can avoid a territorial dispute with the United States over the waters between Alaska and the Chukotsk peninsula, which offshore exploration in the Bering Sea would bring.

The present role of offshore petroleum in solving the problems that beset the Soviet petroleum industry is an easy one to describe. Production of petroleum from the offshore waters is currently only about 1.5% of total Soviet petroleum production. It is therefore unreasonable to think the region can solve any of the production problems currently encountered in the Soviet petroleum industry. However, offshore finds can potentially halt the declining reserves-to-production ratio. This is the context in which to think of the offshore petroleum industry, at least for the next decade. It will not stop the slowdown in production growth caused by the decline in the giant fields of the Tiumen' oil province, nor will it increase the amount of oil for export directly. It will at best, offer Soviet energy planners a possible long-term solution to the problem of declining reserves. This is evident from the geological structures composing the shelf, and it is that to which we now turn.

Geological Survey of Soviet Offshore Waters

The Soviet Union has a continental shelf which is approximately 6 million square kilometers with a depth ranging from 20 to 1,000 meters. The shelf is divided into three regions: first, the intracontinental shelf of the remnant seas, the Caspian, Black and Azov; second, the continental shelf of the Arctic seas and basins, including the Baltic, Barents, Kara, Laptev, East Siberian and Chukotsk seas; third, the continental shelf of the Pacific Far East, which includes the Bering, Okhotsk and Japan seas.

The potential of reserves in the Soviet Union varies as to location and the geological time of the shelf's creation.

Table 3.1 gives a breakdown of the geological time chart by era, period, and epoch.

Table 3.1
Geological Time Scale

Era, Period and Epoch		Duration in millions of years	Millions of years ago
Cenozoic			
Quaternary:			
Recent		0.0	0.0
Pleistocene		2.5	2.5
Tertiary:			
Pliocene		4.5	7.0
Miocene		19.0	26.0
Oligocene		12.0	38.0
Eocene		16.0	54.0
Paleocene		11.0	65.0
Mesozoic *			
Cretaceous		71.0	136.0
Jurassic		54.0	190.0
Triassic		35.0	225.0
Paleozoic **			
Permian		55.0	280.0
Pennsylvanian	***	45.0	325.0
Mississippian	***	20.0	345.0
Devonian		50.0	395.0
Silurian		35.0	430.0
Ordovician		70.0	500.0
Cambrian		70.0	570.0
Precambrian		4030.0	3600.0

* Best prospects on the Soviet shelf
** Second best prospects on the Soviet Shelf
*** Carboniferous structures

Source: Joseph P. Riva, "Soviet Offshore Oil and Gas," *Soviet Oceans Development Committee on Commerce*, (Washington, D.C.: US Government Printing House, 1976).

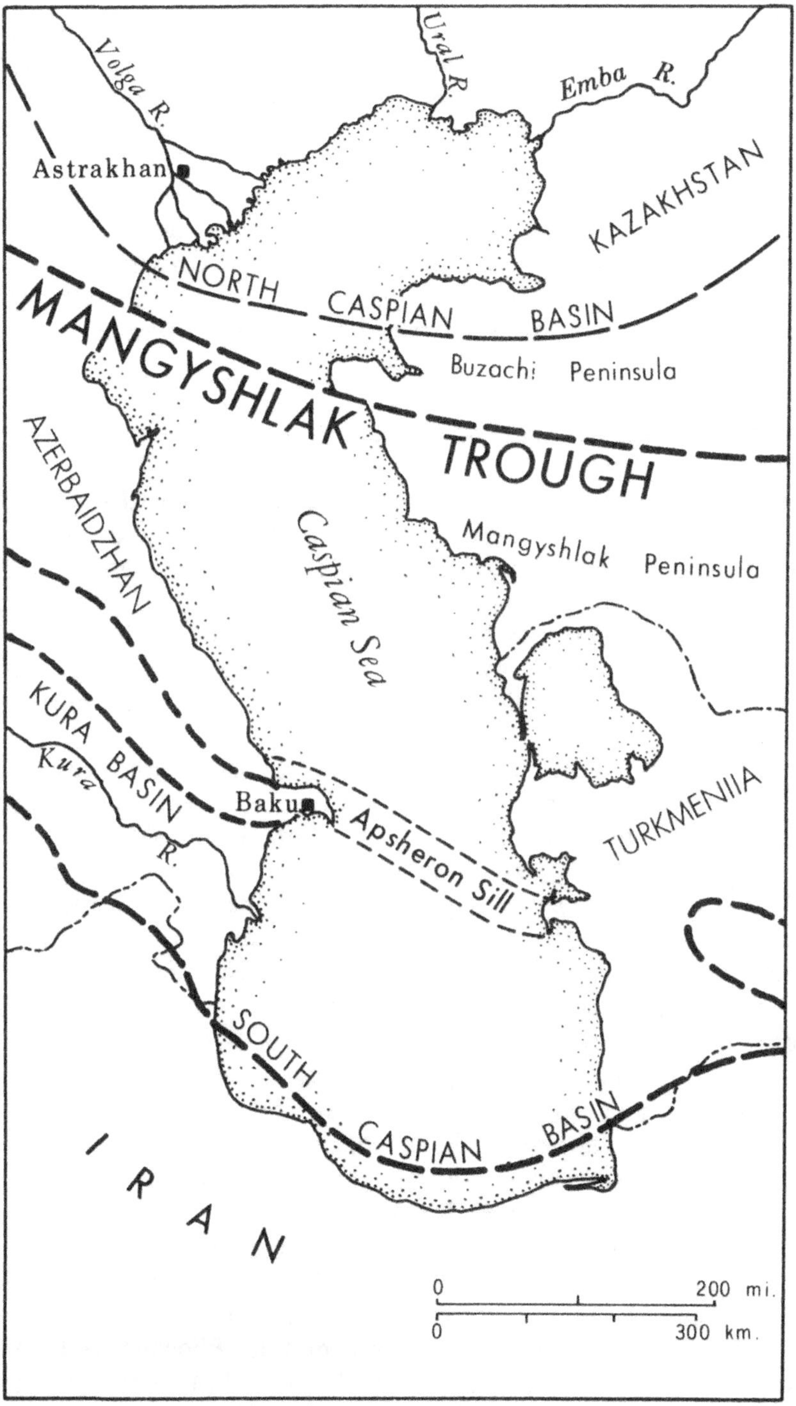

2. The Caspian Sea region

It has been found that commercial hydrocarbons can occur in rocks of Paleozoic, Mesozoic, and Cenozoic ages. On the Soviet shelf the best oil prospects are considered to be at the Mesozoic rock level, with the Paleozoic rock level considered to be the next in importance.[1] The regions of the continental shelf in the Soviet Union all possess some elements of Mesozoic and/or Paleozoic rock ages. Much of the geological knowledge of the Soviet Union's offshore areas, however, is speculative, especially in the Arctic. This should be kept in mind as we discuss the individual basins.

The Intracontinental Shelf of the Remnant Seas

The southern part of the Soviet Union, where the Caspian basin is located, has been fairly well explored and much is known about the geological structure of the region. The northern Caspian depression is composed of a salt dome area that has formed on the Russian Precambrian platform. This platform has been overlaid with thick Paleozoic and Mesozoic rock affording great potential to the northern Caspian Sea.

The middle portion of the Caspian basin, an area that encompasses the area above the Apsheron Sill (an underwater ledge that joins the east and west sides of the Caspian Sea at its narrowest point between the Apsheron and the Cheleken peninsulas), is predominantly covered with Mesozoic and Cenozoic sediments. Here the most potential is from the Jurassic and Cretaceous rocks. The most promising region of the Caspian basin however is the area around the Apsheron Sill, known as the Apsheron-Balkan zone. This sediment is part of the general rock structure of the old Baku fields, and so it is no wonder the region is regarded as having high potential. Some of the largest fields in the Caspian Sea, Neftianye Kamni and the Zhdanov bank are included in this sedimentary basin.

The southern Caspian basin encompassing the area below the Apsheron Sill is seen as nearly as promising as the middle Caspian basin, however, not as well explored. Both have the most prospects from the middle Pliocene sedimentary rock layers. In the southern Caspian Sea, however, the prospects may be even greater than previously thought since the Mesozoic layer rock is almost 20 kilometers thick.[2]

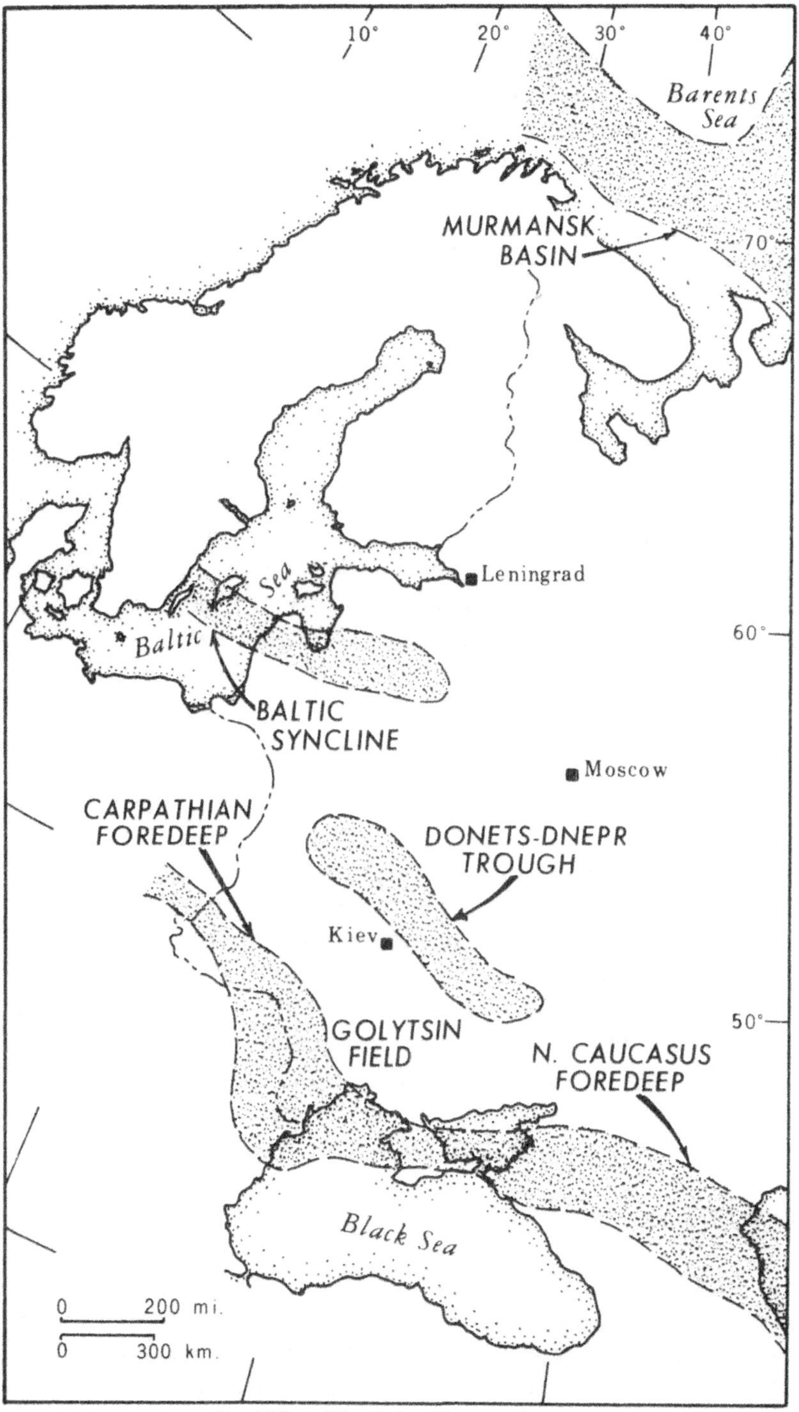

3. The Baltic Syncline

The other two remnant seas, the Black and Azov, have moderate potential with rock layers predominantly of the Cenozoic era. The Black Sea's greatest potential is from an area in the middle of the sea known as the Karkinit deep. Little interest has been expressed in this deep. Instead, efforts have been spent exploring the shallower structures of the Black Sea along Cretaceous and older Tertiary rock.[3] This is probably because of a lack of deep-drilling technology, a problem discussed later.

The Continental Shelf of the Baltic and Arctic Seas and Basins

The continental shelf of the Arctic seas and basins constitutes nearly two-thirds of the Soviet Union's total offshore area. Most of the Arctic shelves are Precambrian rock enhanced by Mesozoic and Cenozoic sedimentation. There are five basins and one trough that make up this area.

The Baltic Syncline is a narrow trough that extends from onshore near Kaliningrad past the island of Gotland in the Baltic and over to Sweden. The Timano-Pechora basin is an extension of the Volga-Ural basin west of the Urals and encompasses the Barents and Pechora seas, the Svalbad plateau, the waters between Franz Joseph island and Novaia Zemlia island, and the northern waters of the Kara Sea. The West Siberian basin encompasses the Soviet Union's giant field of Samotlor and extends offshore into the Kara Sea and south of Novaia Zemlia. The Laptev basin and the East Siberian Sea basin are almost completely offshore, separated by the Novosibirskie islands. The final basin to make up the Arctic offshore region is the Chukotsk Sea basin which is flanked by the Chukotsk peninsula and Alaska.

The Baltic Syncline

The Baltic Syncline is mainly lower-Precambrian rock, not known to be very rich in hydrocarbons. As the trough moves offshore, however, it becomes increasingly thicker with favorable structures. Nearly 60 wells, all of them small, have been discovered onshore near the syncline. With the

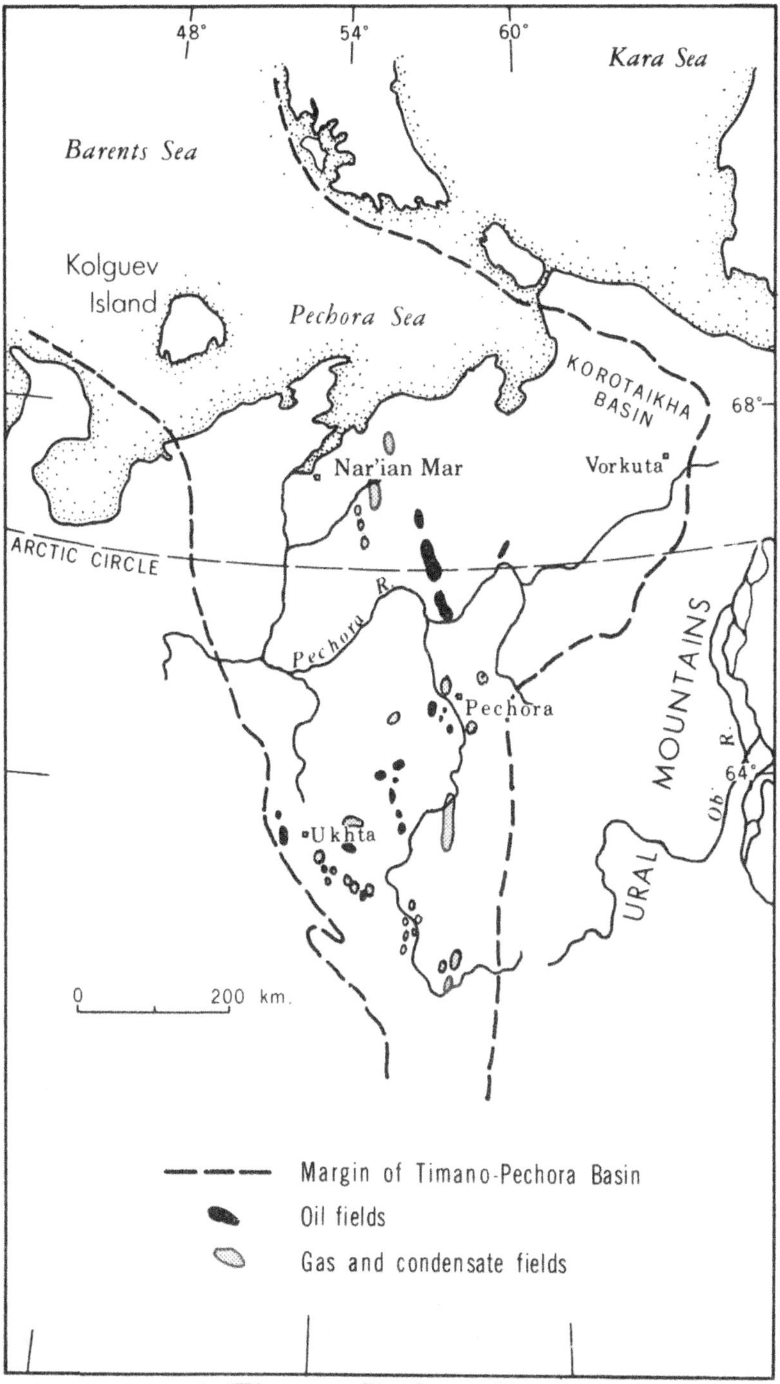

4. Timano-Pechora Basin

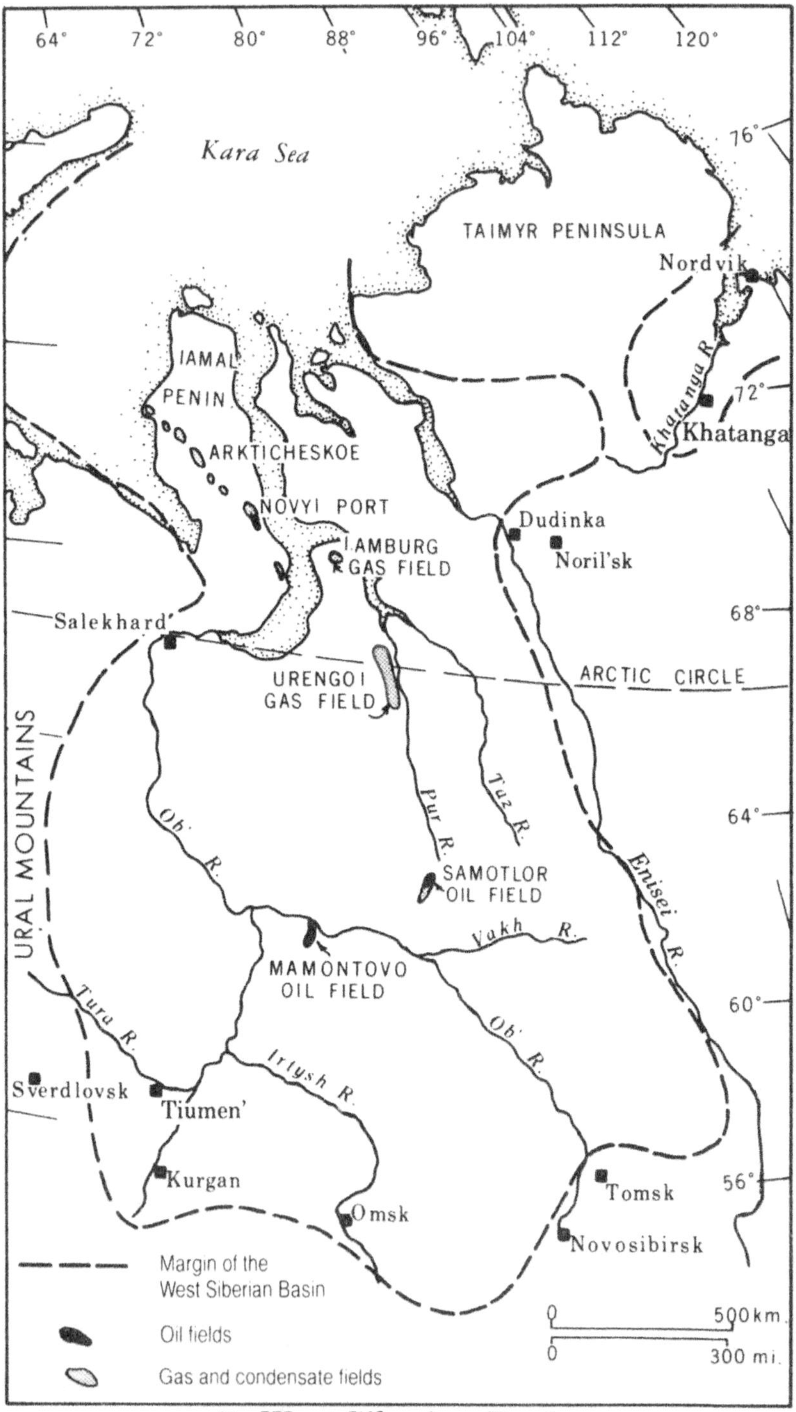

5. West Siberian Basin

increased stratigraphic thickness offshore it is believed that substantial commercial quantities exist.[4]

The Timano-Pechora Basin

The Timano-Pechora basin is believed to be the most hydrocarbon-rich area of Soviet waters. This is in large part based upon the fact that the onshore areas are among the most productive fields in the Soviet Union after the giant fields of the West Siberian basin. The geological structures of the region are not very well known. It is believed, however, that the basin is divided into two parts: the Izma-Pechora platform basin and the Pechora-Kolva Graben, both of which extend northwest and join with the Murmansk basin north of Norway.[5]

The west Barents is believed to be a continuation of the Precambrian structure characteristic of the Murmansk basin north of Norway. The predominance of hydrocarbons is believed to be in the Devonian rock layers that enhance the Precambrian structure beneath. The area of the basin farthest from Novaia Zemlia island near the Norwegian-Soviet marine border at the Murmansk basin is considered to be the most hydrocarbon-rich area of the Arctic. A.A. Meyerhoff claims that this is true because the geologic structures of this region are less complicated by faults and more gentle dips.[6]

The Pechora Sea of the Timano-Pechora basin is also seen as promising. The Mesozoic and Tertiary sediments are cumulatively 15 kilometers thick, and the productive wells along the coast more than likely extend offshore.[7] The shelf around Kolguev island in the Pechora Sea is expected to contain petroleum that is relatively accessible due to the shallow waters of the shelf.

The West Siberian Basin

The offshore region of the West Siberian basin is in the Kara Sea. The region was at one time thought to be extremely oil-rich. This was in large part because the largest oil and gas fields in the world onshore are the, Urengoi and Samotlor fields. The Mesozoic system from Lower Jurassic to the Upper Cretaceous sediments have proven to yield large quantities of hydrocarbons. The structure of the basin, however, begins to change in the north. There is a significant increase in sandstone that is believed to extend offshore. This and the fact that the sea floor has vast uplifts at shallow depths suggest that the Kara Sea is rich in gas, not petroleum.[8]

The extent to which such an assertion is true is based only upon observations of onshore sediments and preliminary seismic surveys along the Kara coast. Ice cover and primitive Soviet seismic equipment have afforded only a scant geologic picture. One alternative theory is that the Kara Sea contains gas at the upper layers of sedimentation (800 to 3,000 meters) and that at lower depths (3,000 to 6,000 meters) commercial quantities of oil exist.[9]

Geological prospecting work, consisting of drilling parametric bore holes, has been done by the Soviet Union in the Kara Sea. The five bore holes drilled so far have revealed that Kara Sea sedimentation was formed between the early Carboniferous and early Cretaceous eras, lending credibility to the theory of the presence of petroleum. Further, studies of cross sections of Mesozoic deposits in the southeastern part of the Kara Sea have been very similar to cross sections of rock of the same age in northwestern Siberia, implying the presence of petroleum.[10]

The Basins of the Laptev, East Siberian and Chukotsk Seas

The next three basins of the Arctic waters--the Laptev Sea, the East Siberian Sea and the Chukotsk Sea are all relatively unexplored. The only offshore work that has been done in the Soviet sector has been for gravity and magnetics; hence, the geological structures of the three basins are only generally known.

The Laptev Sea basin is a distinct tectonic block (tectonic here meaning deformed layers of rocks having a number of folds and faults producing incongruence with the rocks around them). This block is composed of predominantly Lower and Middle Paleozoic sedimentation approximately six kilometers in depth.[11]

There is, however, some controversy over the geological structure of the region, owing to its unexplored nature. If the assertion is that it is a tectonic block composed of primarily Middle Paleozoic sedimentation, then oil and gas prospects would be only fair. The series of folds characteristic of tectonic rock, would suggest some hydrocarbon traps. Yet, as pointed out earlier, Paleozoic rock layers are not believed to be as oil-rich on the Soviet continental shelf as the Mesozoic rock layers. On the other hand, it is believed that the Laptev Sea basin is composed of carboniferous (also Middle Paleozoic of the Pennsylvanian and Mississippian eras) through to Cretaceous (Upper Mesozoic) rock layers. If this is the case, the prospect of finding hydrocarbons would likely be greatly increased.[12]

The East Siberian Sea and Chukotsk Sea basins lack both analysis and speculation to any extent about their geologic structures. For the most part the region is composed of Upper Paleozoic, Mesozoic, and Cenozoic rock layers and is refered to only as "...prospectively oil and gas bearing" (an obvious and ambiguous conclusion). The Hyperborean platform which contains most of the hydrocarbon deposits in northern Alaska is also thought to underlie both seas. If this is the case, then the region would have excellent potential, both from a geological and a developmental point of view since the waters in both seas are quite shallow, averaging about 50 meters.[13]

The Far Eastern Waters

The waters of the Soviet Far East are comprised of nine basins. These are from north to south: the Anadyr basin, which includes the Anadyr gulf and extends into American waters near St. Lawrence island off the coast of Alaska; the Khatyrka basin, which includes the southern portion of the Anadyr gulf and the northern portion of the Bering Sea; the basins of Kamchatka, which encompass the offshore area of the Kamchatka peninsula, most of the Bering Sea, and the northwestern tip of the Pacific Ocean; the northern Okhotsk basin, which includes the Shelikhova gulf between the Kamchatka peninsula and the east coast of the Soviet Union and the northern portion of the Sea of Okhotsk; and the north Sakhalin basin, which includes most of Sakhalin island and the waters off its eastern coast. The next four basins, the Pogranichnyi basin, the Terpeniia gulf basin, the Aniva Gulf basin, and the Tatar Strait basin, are rather small in comparison to the other Far Eastern basins; they extend from the lower portion of the north Sakhalin basin to

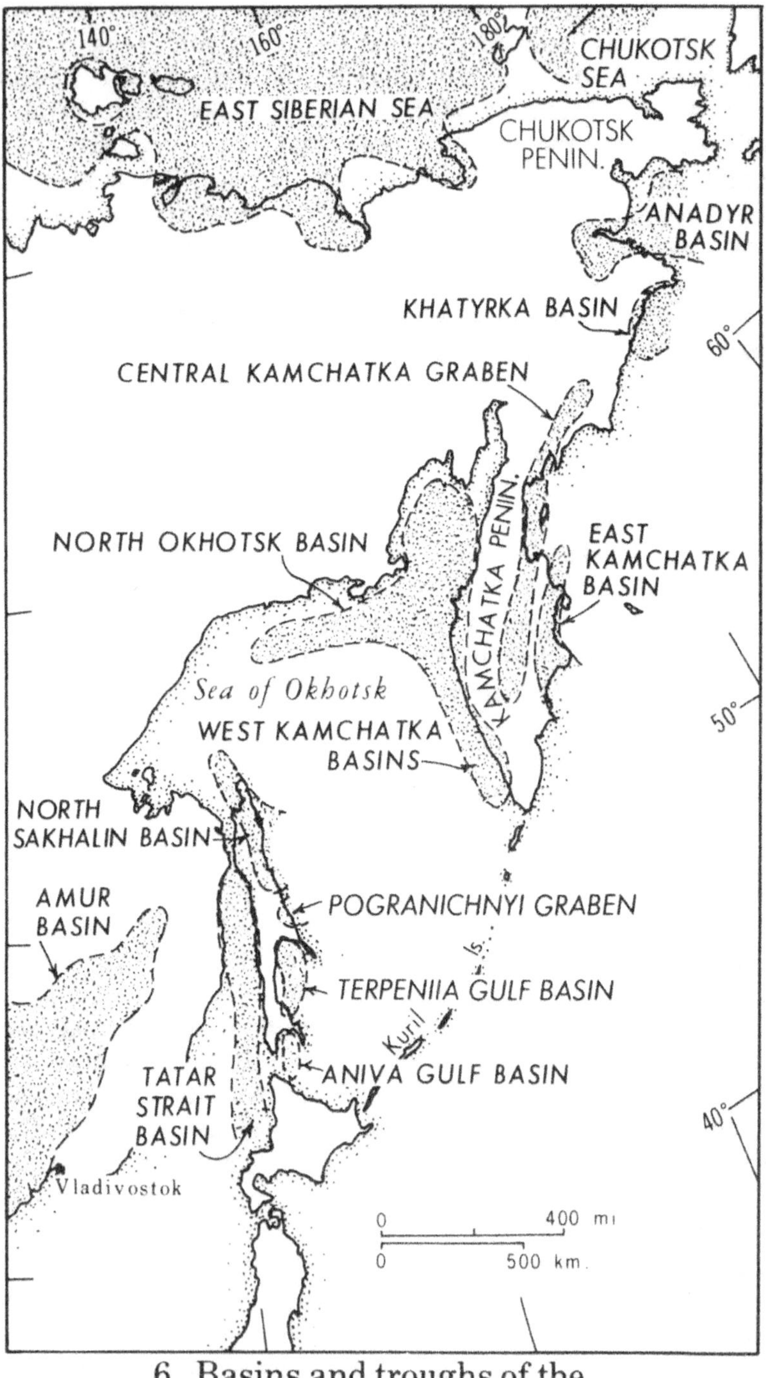

6. Basins and troughs of the Soviet Far Eastern waters

Hokkaido island of northern Japan through to the waters off of Vladivostok.

The Anadyr and Khatyrka Basins

The Anadyr basin is underlaid with Tertiary rock, predominantly of the Miocene and late Oligocene and Eocene eras. The only areas that have been extensively explored, however, are the northern regions of the Bering Sea. As a result the geological structure is relatively unknown. One thing which seems certain, despite the lack of seismic work in the area, is that the region is predominantly a gas trap with some gas condensate. The porous and permeable sandstone sediments discovered in the Cenozoic sediments onshore would suggest this.[14]

The geological structure of the Khatyrka basin is thought to be generally similar to that of the Anadyr basin. Oil and gas seeps have been found from the Eocene and Oligocene rock strata onshore, and it is speculated similar structures extend offshore. The promise of commercial petroleum is slight. The area holds the most promise for gas.[15]

The Kamchatka Basins

There are four small basins comprising the Kamchatka region: the east Kamchatka basin, the central Kamchatka trough, the Priokhotsk basin, and, the Parapol'Poliana basin. The region is a complex series of folds and volcanic rock owing to its proximity to the oceanic crust. Late Cretaceous through Pliocene sediments are believed to exist offshore. A number of authors believe there may be potential for near-commercial quantities of oil and gas, but nothing yet to date substantiates this belief.

The North Okhotsk and North Sakhalin Basins

The geology of the north Okhotsk basin has not been explored due to the ice cover, remoteness, and near lack of any hydrocarbon deposits onshore. It is believed the region, if developed, would yield gas predominantly.

The north Sakhalin basin is the most explored region in the Far Eastern waters. The discovery of commercial oil on Sakhalin is largely responsible for this. The largest fields onshore are Okha, Vostochno-Ekhabi, Kolendo, Ekhabi and Tungor; offshore the largest field is the Odoptu structure. Most of the fields are underlaid by late and middle Miocene rock layers believed to extend up from the Sea of Japan and the Tatar strait basin.[16]

The offshore field, Odoptu, is believed to contain very large quantities of oil and gas. It is speculated the offshore waters of Sakhalin will yield more hydrocarbons than onshore. This is substantiated by the fact that the offshore geological structures are larger and gentler than onshore and hence less fault-ridden.[17] The field is not the largest oil field discovered in eastern Asia, however, as both the Soviet Union and Japan claimed when they began jointly to develop the region. Geologists have put the estimated potential of the Daqing field in northeastern China at more than 5 bbl.[18] The Odoptu field is nonetheless important because it is the only commercially productive offshore field in the entire region whose demand for petroleum is growing yearly.

The Pogranichnyi, Terpeniia Gulf, Aniva Gulf and Tatar Strait Basins

The Pogranichnyi basin is thought to be an extension of the north Sakhalin basin. The area both on and offshore is composed primarily of Miocene sediments overlaying Cretaceous rock layers. The region is not thought to contain commercial quantities of oil and gas. It does have one producing onshore well, Okruzhnie.[19]

The Terpeniia gulf and Aniva gulf basins are composed mainly of Cretaceous rock layers overlaid with Miocene sedimentation. These layers are thought to contain most of the hydrocarbons of these two areas. The Terpeniia gulf, however, has a lack of good reservoirs, a characteristic of the Cretaceous rock in the Far East. The Aniva gulf basin has produced only one commercial gas well, Vostochno-Lugovsk which is both on and offshore.[20]

The Tatar Strait basin has one of the poorest ratings for petroleum in the Far East. Only slight oil and gas seeps have come from the Miocene through Paleocene rock layers. The basin is not thought to have very much potential because of the relatively small finds on Hokkaido island, which is on the same geological structure.[21]

Summary of the Offshore Region's Potential

In May of 1985, the Offshore Technology Conference (OTC) held a series of seminars, one of which ranked the hydrocarbon potential of the world's offshore and partly onshore-offshore reserves. According to the study, there are 799 sedimentary provinces in the world which possibly contain hydrocarbons; 481 of these are located offshore or partly offshore. Of these 481, 8 are considered to have excellent hydrocarbon potential, 34 good-to-excellent, 42 good, 72 fair-to-good, 100 fair, 80 poor-to-fair, and 145 poor.[22] The 145 hydrocarbon provinces rated as poor were not listed by the conference. Of those listed, 25 hydrocarbon provinces were in the Soviet Union or were shelves in waters the Soviet Union shares with other countries. In the excellent potential category the Soviet Union had 2, in the good-to-excellent category 6, and 2 in the good category, 5 in the fair-to-good, 5 rated fair, and 5 rated poor-to-fair. Table 3.2 shows these rankings:

Table 3.2
Soviet Offshore Hydrocarbon Potential

Province	Area	Onshore/ Offshore	Offshore
Excellent Potential			
North Slope	Alaska Bering Sea	X	-
North Sea Northern	Norway Murmansk basin	-	X
Good-to-Excellent Potential			
Caspian South	Caspian Sea USSR/Iran	X	-

Continued on next page

Table 3.2 continued

Province	Area	Onshore/ Offshore	Offshore
Mangyshlak	Caspian Sea USSR south	X	-
Caucasus North	Caspian Sea USSR south	X	-
Kara Sea	USSR Arctic	-	X
Pechora	USSR Arctic	X	-
Barents West	USSR/ Norway	-	X
Good Potential			
Barents East	USSR Arctic	-	X
Black Sea	Rumania/ USSR	-	X
Fair-to-Good Potential			
Sakhalin North	USSR East	X	-
Hope	Alaska/USSR	-	X
St. Matthew	Alaska Bering Sea	-	X
Tatarskii	Japan/USSR	X	
Baltic	Poland/ Sweden/USSR	-	X

Continued on next page

Table 3.2 continued

Province	Area	Onshore/ Offshore	Offshore
Fair Potential			
Anadyr	USSR	X	-
Navarin	Alaska/USSR	X	-
Norton	Alaska/USSR	X	-
Kotzebue	Alaska/USSR	X	-
Terenyi	USSR East	X	-
Poor-to-Fair Potential			
East Siberia Sea	USSR Arctic	X	-
Barents Fan	USSR/Norway	-	X
Central Kamchatka	USSR East	X	-
Prins Karl	Norway (Spitsbergen)/USSR	-	X
South Okhotsk	USSR East	-	X

Source: B. St. John, "Offshore Exploration Potential," *1985 Offshore Technology Conference* (May 6-9, 1985), Houston, Texas.

The regions the Soviet Union is developing and will further develop in the future are those predominantly petroleum-rich. Those believed to contain large quantities of gas and condensate therefore, while being rated by the OTC, will not be of major concern.

Of the provinces listed in the OTC survey, all of those ranked from good potential to excellent potential have geologic structures suggesting the presence of petroleum. The only exception is the Kara Sea where the expectation of commercial quantities of gas has placed it in the second-highest category. The ongoing debate over the Kara Sea's hydrocarbon content and the fact that the sea is quite shallow means it may become a Soviet area of exploration anyway. Some of the provinces listed by the OTC as poor-to-fair or fair-to-good are predominantly gas traps or do not contain commercial quantities of petroleum, as is known from the above geological survey of the Soviet offshore waters. Those provinces are: St. Matthew, located in the Anadyr basin (its rating is probably attributable to gas finds on the Alaska side of the island); Tatarskii located between the island of Hokkaido and the Soviet Union's southeastern coast (a rating also due to the predominance of gas); and Norton, also located in the Anadyr basin.

The area with the most offshore potential, as shown in Table 3.2, is the Timano-Pechora region. The Murmansk basin is an extension of this region. It is rated in the excellent category. Similarly, the Barents West and the Pechora Sea are rated in the good-to-excellent category and represent the core of the Soviet Union's efforts in the Arctic waters. The other northern region ranked in the excellent category is the Bering Sea. So far, however, most of the petroleum discoveries in this region have been within American waters. The areas closest to the Soviet Union are believed to contain primarily gas. Finally, the Caspian region is rated in the good-to-excellent category. The more temperate zones there have allowed for more extensive seismological surveys to be undertaken. There may, therefore, be a bias in the favourable assessment of this region. Nonetheless, the efforts expended so far attest to the Caspian region's crucial role in the Soviet offshore petroleum industry.

There are in all, therefore, 22 potentially petroleum-rich provinces surrounding the Soviet Union. The most promising are the Caspian Sea region (the areas defined by the north, middle and southern portions of the Caspian basin), the Baltic Sea (the area defined by the Baltic syncline), the

Barents and Pechora seas (those areas defined by the Timano-Pechora basin), and the waters off north Sakhalin island (defined by the north Sakhalin basin).

The reserve capacity of the offshore regions is substantial. Given full organizational and technological capability, the offshore regions soon could reverse the declining trend in the present reserves-to-production ratio. The question is if the Soviet Union can overcome its organizational and technological constraints. Knowing petroleum is there geologically is one thing, being able to access it in large quantities quite another. The Soviet Union has consequently taken steps to ensure its capability to exploit the offshore deposits.

Notes

1. Joseph P. Riva, "Soviet Offshore Oil and Gas," Soviet Oceans Development Committee on Commerce (Washington D.C., October 1976), p. 482.
2. Joseph P. Riva (1976), pp. 481-482.
3. Joseph P. Riva (1976), p. 483.
4. Little information on the Baltic subsea reserves is known. When oil was found offshore at Kaliningrad it was believed that the find was insignificant. A study undertaken by N. A. Eremenko, Ya. P. Malovitsky, I.S. Gramberg, and L.I. Labedev reported in The American Association of Petroleum Geologists Bulletin (vol. 57, No. 2, Feb. 1973, pp. 240-241), suggests that the petroleum reserves of the region increase seaward. See Joseph P. Riva (1976), p. 483.
5. A.A. Meyerhoff (1980), p. 50.
6. A.A. Meyerhoff (1980), p. 50.
7. Joseph P. Riva (1976), p. 483.
8. See Ermenko, N.A., Ovanesov, G.P. and Semenovich, V.V. (1972), p. 1716, cited in Joseph P. Riva (1976), p. 484.
9. A.A. Meyerhoff (1980), pp. 105 - 108.
10. I.S. Gramberg, I.V. Shkola, Sovetskaia geologia, No. 1, Jan. 1985, pp. 95-98, cited in JPRS UES-85-006-66A.
11. Joseph P. Riva (1976), p.484.
12. The belief that the Laptev basin is predominantly composed of Middle Paleozoic sedimentation is based on the 1972 work of Eremenko cited in Riva (1976). Meyerhoff,

however, cites a later work: G. Kh. Dikenshtyn, I.M. Aiyev, G.A. Arshevskii, V.A. Kirov, A.E. Kontorovich, S.P. Maksimov, V.D. Nalivkin, I.I. Nesterov, N.S. Sabaneiev, V.V. Semenovich and Yu. N. Shbemberger, Neftegazonosnye provintsii SSSR (Moscow: Izd. Nedra, 1977), which claims that the region is composed of rock layers from Middle Paleozoic to Upper Mesozoic. Clearly, the geologic knowledge of the region is unclear.

13. N.A. Eremenko et. al., cited in Joseph P. Riva (1976), p. 484.

14. Both Riva and Meyerhoff conclude this. See Riva (1976), p. 485, and Meyerhoff (1980), p. 120.

15. A.A. Meyerhoff (1980), p. 120.

16. A.A. Meyerhoff (1980), Table 3, p. 132.

17. Meyerhoff claims this to be the case, but stresses the tenuousness of the assertion. See A.A. Meyerhoff (1980), p. 143.

18. A.A. Meyerhoff (1980), p. 143.

19. A.A. Meyerhoff (1980), p. 143.

20. A.A. Meyerhoff (1980), p. 144.

21. A.A. Meyerhoff (1980), pp. 144-145.

22. B. St. John (Primary Fuels, Inc.), "Offshore Exploration Potential," 1985 Offshore Technology Conference (May 6-9, 1985), Houston, Texas, vol. 1, p. 307.

4

Soviet Offshore Petroleum Exploration and Development

The purpose of this chapter is threefold: first, it will detail and assess the administrative reorganizations that have occurred in the Soviet offshore petroleum industry; second, it will describe Soviet production and exploration activities to date in the offshore waters; third, it will detail the needs of the Soviet offshore industry, particularly as they apply to deep-water technology and offshore Arctic petroleum development. The latter will involve a discussion of the type of equipment required for offshore exploration and production and the characteristics of the Western partners involved. In order to address the questions posed by the practical ability of the Soviet Union to undertake offshore exploration and development, we shall deal with the regions as defined by seas rather than by basins. This is also the way administrative and press reports deal with the offshore regions.

Administration of the Offshore Petroleum Industry

The administration of the offshore industry has evolved from the Soviet Union's experience to date in offshore petroleum production and knowledge of potential reserves.

There are only two areas actually producing any oil, the Caspian and Okhotsk seas, for two reasons--lack of technological capabilities, and the need to meet regional demands. This has meant that planners have explored and developed the offshore regions closest to the Soviet Union's industrial centers.

The first site of Soviet offshore oil exploration and development was the Caspian Sea because of its proximity to the Baku region, the Soviet Union's first large-scale oil province. Most of the development in this region was on a small scale until 1960 when the Soviet Union decided to invest substantially in offshore equipment for the region, although generally placed only in the shallow waters of the Caspian Sea along the Apsheron Sill. Large-scale, deep-water development in the region did not begin until much later, in the 1970s.

The next site emphasised for exploration was the northern waters of the Sea of Okhotsk. Slant drilling had begun there in 1971 off the eastern coast of Sakhalin island to tap the near shore resources of the island's already explored and developed reserves. The reason for development here was to service the growing demands of the Ussuri and lower Amur valleys. According to one source, petroleum was in such short supply in the region that oil had to be shipped in by tank car from the Volga-Ural fields and West Siberia.[1]

In both the Caspian and Okhotsk seas production of petroleum was attributable to two factors: first, the need to meet local demands, an objective facilitated by already existing onshore petroleum producing sites; and second, the relatively easy technological challenge of drilling in no more than 25 meters of water.

As offshore petroleum became more of a concern in the mid-to-late 1970s, planners began to take a more systematic approach to exploration and development. Although servicing local demands and the proximity of existing onshore petroleum facilities remained important considerations, it became more important to develop the regions for long-term purposes. The need to organize the industry under one of the three ministries--Petroleum, Gas, or Geology--became essential, as did the need to formulate long-term strategies for exploration, assessment and technical requirements.

Until 1978, offshore exploration and development was *ad hoc* and reactive. The result was that offshore work was divided among the ministries of Petroleum, Gas and Geology as well as the State Mining and Technology Control Agency, Gostekhnadzor. Each body pursued its own development strategy with minimal consultation. The result was that little was accomplished beyond the shallow drilling in the Caspian Sea and the slant drilling in the Sea of Okhotsk.[2]

The Ministry of Geology was responsible for most of the offshore work outside the Caspian Sea. The problem was that most of the experience being gained was in the Caspian Sea, the responsibility of several subdivisions of the Ministry of Petroleum. The problem was even more complex, however, because the Ministry of Geology would only undertake the geophysical work, leaving the responsibility for prospecting and development plans to the other two ministries.

The problem was particularly pronounced on the research and technical development side. Several different research institutes of the USSR Academy of Sciences, other schools and other ministries and agencies were undertaking the study of various problems related to offshore exploration and development. The result was a lack of coordination and accountability, and a tendency among many of the organizations to treat offshore work as a sideline activity to which small and poorly equipped teams were assigned.[3]

In an effort to overcome these difficulties, it was decided to concentrate all offshore activity in the Ministry of Gas. It seems that Soviet planners realized that the costs of exploration and development were high enough without driving them even higher by bureaucratic rivalry and technological duplication. The Ministry of Petroleum has estimated that exploration costs in the Caspian Sea alone are nearly 20% of the total costs of offshore development. This estimate was made in 1985, well after the reorganization took place. One can imagine how much higher the estimate would have been before restructuring took place.[4]

It was decided that research work would be organized along a three-level system, typical of many branches of Soviet industrial production. The levels would be an all-union association, a production association and an enterprise level.[5] The intent was to avoid the disorganization problems encountered in the past. The reorganization was undertaken in earnest and followed the type of economic restructuring the Soviet Union has pursued when developing other industries. For example, the creation of the ob"edineniia along three-tier lines was to increase the flow of information between ministries and enterprises and create economies of scale in research and development. The reorganization of research and development stemmed from an understanding by planners of the need to improve drastically the offshore industry's technical base. I. Khvedchuk, the chief geologist of the All-Union Marine Geology Research and Production's Pacific Ocean Offshore Expedition, pointed out in 1978 there were not enough "...specialized geophysical vessels, floating drill rigs, and other up-to-date devices" to develop the offshore waters. This includes improving existing technology, importing new equipment and designing new rigs for deep-water exploration.[6] Presumably, this was to be facilitated by concentrating responsibility under one ministry.

The concentration of the administration of the offshore industry in the Ministry of Gas should not be viewed as a concession to the exploration of gas at the expense of petroleum. The concentration was both political and pragmatic because of the relative efficiency displayed by the Ministry of Gas over the previous decade and its increasing political clout. Press reports on the concentration of the offshore industry under the Ministry of Gas make reference to the exploitation of petroleum (*neft'*), not oil and gas (*neft' i gaz*).[7] The emphasis of this reorganization is on bureaucratic concentration in a ministry which has shown itself to be more aggressive in both exploratory drilling and organizational ability than either the Ministry of Petroleum or the Ministry of Geology.

The political reasons for the concentration of offshore work in the Ministry of Gas may be a result of infighting during the 1970s between the petroleum ministry and geology ministry, and the petroleum ministry's lack of attention to exploratory drilling.

In 1974, oil minister Valentin Shashin criticized the Ministry of Geology for not drilling enough exploratory wells in prospective new provinces. Shashin suggested that the Caspian depression, East Siberia, and "...the seas and shelves of the oceans which have enormous potential for growth of reserves" may be areas to explore.[8] The implication being possibly that the offshore work should be taken away from the Ministry of Geology and concentrated in the petroleum ministry. It was only shortly after this, however, that the Ministry of Petroleum came under criticism for the declining output of petroleum from the Tiumen' oil province. With the emphasis on petroleum production over development drilling through the 1970s, the Ministry of Petroleum probably had its hands full reorganizing to meet the output quotas which were continually being raised.

It is quite possible, therefore, that the responsibility for offshore petroleum exploration fell upon the Ministry of Gas because it has increased its exploratory drilling at a dramatic pace over the last decade and a half. There is an indication the Ministry of Gas will also be responsible for development of the offshore reserves. The reason is because, at present, Kaspmorneftеprom, the agency responsible for offshore oil production in the Caspian Sea, is part of the Ministry of Gas.[9]

It is probably better to reserve judgment about who will be given the larger and more prestigious job of offshore development. The death of the Minister of Gas, S.A. Orudzhev in 1981 and the replacement of the First Deputy Minister of Gas, Iu.V. Zaitsev in 1983, have apparently shaken the offshore operations in the Ministry of Gas. There may be more changes. These developments and the recent political emphasis bestowed upon the Ministry of Petroleum by Gorbachev in the last party congress suggest the possibility of more administrative changes once offshore exploration becomes better organized.[10]

While it is probably true to say the administration of offshore production in large quantities from the Arctic or the Sea of Okhotsk is not possible to ascertain at this time, the administration of the shelf's exploration would seem to be well organized and potentially quite successful. The centralizing of offshore petroleum exploration in the Ministry of Gas should be viewed as a hopeful sign for three reasons.

First, the current issue for offshore petroleum planners is exploring the reserves in the offshore regions, not preparation for production. In the field of exploration, the Ministry of Gas, with its successful exploration over the past decade in western and eastern Siberia, has positioned itself much more strategically than either the ministries of petroleum or geology. Both of the latter ministries have been beset with problems from the late 1970s to early 1980s, in large part stemming from the draconian output quotas placed on petroleum during these years. As a result, it was probably felt that the Ministry of Gas was better positioned to fulfill the task of exploration.

Second, since the build-up of the Soviet gas industry in the early 1970s and the construction of the Soviet-Western European gas pipeline, the Ministry of Gas has had considerable experience with the importation of Western technology. During the Carter administration in the United States from 1977-1981, the Soviet Union began to experience problems obtaining American technology imports and with the licensing of American technology to Japanese rigs in the Sea of Okhotsk. The subsequent enlisting of numerous Western European partners by the Soviet Union may have been part of the reason for concentrating the administration of offshore exploration in the gas ministry. The ministry with its recent contacts in Western Europe for energy technology was probably well positioned. Further, it was the gas ministry which dealt with American sanctions more than either the ministries of petroleum or geology. To get around American sanctions the gas ministry cultivated a variety of technology sources. The amount of Western technology necessary to develop the offshore waters suggests the offshore administration is well placed in the Ministry of Gas.

Finally, with the decline in the price of gas even relative to petroleum and the decreasing demand for Soviet gas in Western Europe, the Ministry of Gas is probably in a better position to administer the offshore regions than the ministries of petroleum or geology. The Ministry of Gas is probably anxious to maintain its successful image gained over the past decade, and offshore petroleum exploration is one way of doing this.[11]

It is important now to discuss the Soviet Union's experience to date in offshore petroleum exploration and production. Knowledge of the geological composition of the shelf is an excellent guide to predicting actions and assessing the reasons for a particular activity in the offshore waters. The administrative centralization and the apparent vying for position as to who will be responsible for development of the offshore reserves suggest that some serious considerations are being given to the region and its development is not simply *ad hoc* and reactive.

General Offshore Petroleum Production and Perspectives of Analysis

In 1986, total offshore production of petroleum was about 200 tbd, all of it from the Caspian Sea and primarily from three large fields. The production from the Sea of Okhotsk has been halted and in the other regions, only exploration is underway. Gas has been produced in small quantities from the Black Sea, and by all reports production there had ceased by early 1980.

Trends in offshore production are shown in Table 4.1. Production had its largest surge between 1960 and 1970, rising from 143.5 tbd in 1960 to 228.6 tbd in 1965 and finally peaking at 258.0 tbd in 1970. Since this time there has been a steady decline in the rate of output, with only one minor increase between 1972 and 1973.

This does not paint a very optimistic picture for the offshore industry. With production declining, fulfilling less than 2% of Soviet petroleum output, one could tend to ignore the offshore industry in the Soviet Union. The fact is, however, that until quite recently the Soviet Union was only able to explore and produce in the warm and shallow waters of the Caspian Sea. Most of its offshore reserves are estimated to lie in fields yet to be adequately explored. The Soviet Union has taken an interest in these fields as is evident through their numerous seismological surveys in the Barents Sea and the north and south basins of the Caspian Sea, their administrative centralization in 1978, the purchase of Western equipment in recent years, and their emphasis on domestic offshore technology.

What emerges from a discussion of the historical development of the Soviet offshore industry is the conclusion the Soviet Union realized how much preparation needed to be done in an organized and systematic fashion before it could reasonably hope to raise production of the offshore fields to, say, the United States' level of 14%. The Soviet Union's situation is more akin to Norway's. Most of its potential fields lie in similar waters; hence, the same preparation is required. It took the Norwegian industry, with cooperation from the international petroleum industry, over ten years to acquire enough experience to operate and produce petroleum from a major field. This does not include the preparation involving seismic surveys and exploratory drilling.[12]

Table 4.1
USSR Offshore Petroleum Production, 1960-1986

Year	Production (thousand barrels/day)	% of Total	Year	Production (thousand barrels/day)	% of Total
1960	143.46	4.8	1975	228.00	2.3
1965	228.58	4.7	1976	220.00	2.1
1966	238.44	4.5	1977	205.00	1.9
1967	247.88	4.3	1978	200.00	1.7
1968	252.44	4.1	1979	190.00	1.6
1969	255.00	3.9	1980	197.00	1.6
1970	258.00	3.7	1981	196.00	1.6
1971	250.00	3.4	1982	175.00	1.4
1972	236.00	3.0	1983	175.00	1.4
1973	240.00	2.8	1984	180.00	1.5
1974	231.00	2.6	1985	190.00	1.5
			1986	200.00	1.5

Sources: 1960-1970: *Offshore* (June 20, 1977), p.204; 1975-1981: *Offshore* (June 20, 1982), p.145; 1982: *Offshore* (June 20, 1983), p.108; 1983: *Oil and Gas Journal* (June 4, 1984); 1984-1985: *Offshore* (May, 1986), p. 110; 1986: *Petroleum Economist* (May, 1986), p.188.

The perspective of analysis concerns what the Soviet Union has learned over the past six decades of dabbling in the offshore petroleum industry. There is no doubt that interest has increased in the offshore industry. One explanation is the Soviet Union has realized the role of offshore petroleum in its energy industry. What it is about to find out is the cost of casting this role.

The Caspian Sea

The Caspian Sea is located in the Caspian basin where six oil and gas provinces join. These six provinces are among the most productive in the Soviet Union after those of the Timano-Pechora basin and the West Siberian basin. They are the oil province of Azerbaidzhan, south Dagestan, eastern Predkavkaz, the Caspian depression, Mangyshlak (Kazakhstan), and western Turkmeniia. The historical development of the offshore industry in the Caspian Sea can be divided into four distinct periods.

Early Development

In the first period, between 1904 and 1922, oil was produced only on the Artem islands, two islands north and south off the Apsheron peninsula located on the Apsheron Sill. Offshore work on the islands began in earnest in 1909 under the direction of the engineer, P.N. Potatskov. Here wells were drilled on the filled-in bay of Bibi-Ezhbata. Little was yielded here and work was halted shortly thereafter as a result of the political turmoil which ensued prior to and after the revolution.[13]

In the second period, between 1923 and 1948, work began again on the Artem islands under the direction of S.M. Kirov, a full two years after the Soviet pacification of the area. The period has certain important developments which were to characterize the early offshore industry in the Caspian. The first well to yield oil in this period was on another filled-in bay named Bukhtoe Il'ich. In this period, five wells were under production. Work soon began, however, to move away from the filled-in bays to north of the Artem islands. In 1934, Artemneft' was formed. It was an enterprise association (*trest*) to ensure the exploitation of oil and gas deposits in the northern waters off the Artem islands. It was at this time that the Soviet Union first constructed metal platforms to drill from the shallow waters of the sill.[14]

The early platforms were no more than large metal trestles, crudely constructed onshore, then dropped offshore in a hit-and-miss fashion. What is significant is that because

7. Offshore developments in the Caspian Sea

of technological constraints, the Soviet Union only slowly expanded along the shallow Apsheron Sill and only as regional demand for oil called for it. This type of development and exploration changed only as recently as 1980. The technology has increased in sophistication, and the strategies of development have altered over the recent years.

The Discovery of Neftianye Kamni

The third period of development is between 1949 and 1962. In this time seven deposits were discovered at Giurgian-more, and the first giant region--Neftianye Kamni. It was with the discovery of Neftianye Kamni in the latter part of 1949 that offshore production began to increase at a rapid pace. This field had an estimated initial reserve of 1.2 bbl and was seen as the hope for the whole region.[15] Production was not a particularly difficult since the deepest well at Neftianye Kamni was only 35 meters and the shallowest 6 meters.[16] Between 1949 and 1986 over 1,200 wells were drilled here.[17]

From this time on numerous finds along the sill yielded good test-pays yet none were of the magnitude of Neftianye Kamni. Most of the Soviet Union's efforts were concentrated on this field. By 1970 it was producing over 60% of all offshore production at 140 tbd.[18]

Dwindling Production

The fourth period dates from 1963 to 1975. The Soviet Union expanded exploration along the sill again, finding 24 deposits at Sangachali-more, Duvannyi-more, Bulla island, Garasu, and the Bakhar field to the southeast of the sill.[19] Of all of these, Garasu with a test flow of 1,825 bd was the most promising, yet little was done after the initial wildcatting.[20] There seemed to be little interest expressed in these reserves despite their promise, perhaps because there was no real plan for the development of the offshore waters. As long as oil was easy to reach in the shallow waters of the Apsheron Sill there was no real need to explore further. Yet by 1975, petroleum production from the region began to fall off dramatically. This occurred for two reasons.

First, on the Apsheron Sill surrounding Neftianye Kamni and the area southwest of Baku where Duvannyi-more and Bulla island are located, the major facilitating factor was the trestle causeways that serviced the areas. The trestle causeways failed to reach further out, which explains the decline in petroleum production.[21]

Second, Neftianye Kamni was beginning to reach its full capacity by the mid-1970s. From an estimated reserve of 1.2 bbl in 1949, it was estimated the field's wells had cumulatively produced 1.1 bbl by 1986.[22] While it is possible the 1949 estimates were conservative, the fact Neftianye Kamni has not increased production since 1970 means hopes for new reserves at the site are unlikely.

The concern over production from the offshore fields became more acute by 1975, when it was discovered that offshore production, while insignificant to overall oil production in the Soviet Union, was by far the major contributor to Azerbaidzhan's petroleum output. Between 1970 (the high point of Neftianye Kamni's output) and 1980, total Azerbaidzhan petroleum production fell from 21 mty to 14 mty. Over 70% of the petroleum output came from the offshore fields.[23]

The Contemporary Offshore Situation in the Caspian Sea

Present offshore developments date from 1975 to the present, and represent the Soviet Union's effort to systematically organize itself for offshore exploration and production. It was clear to Soviet planners that if offshore petroleum was to become a reliable industry, deep-water exploration and technology would be necessary. Further, it was necessary to stem the declining production in Azerbaidzhan. Efforts were made, therefore, to discover as much petroleum as possible along and off the Apsheron Sill. In 1977 a field named "28th of April", was discovered about 20 kilometers southeast of Neftianye Kamni. This field, while not having the estimated reserve capacity of Neftianye Kamni's 1.2 bbl, was nonetheless seen as the major hope to compensate for the decline in the Caspian Sea production.

Drilling in the 28th of April field began early in 1977 on the southwest bank. By 1979 eight wells had been drilled. In 1980, production began from the field as did increased

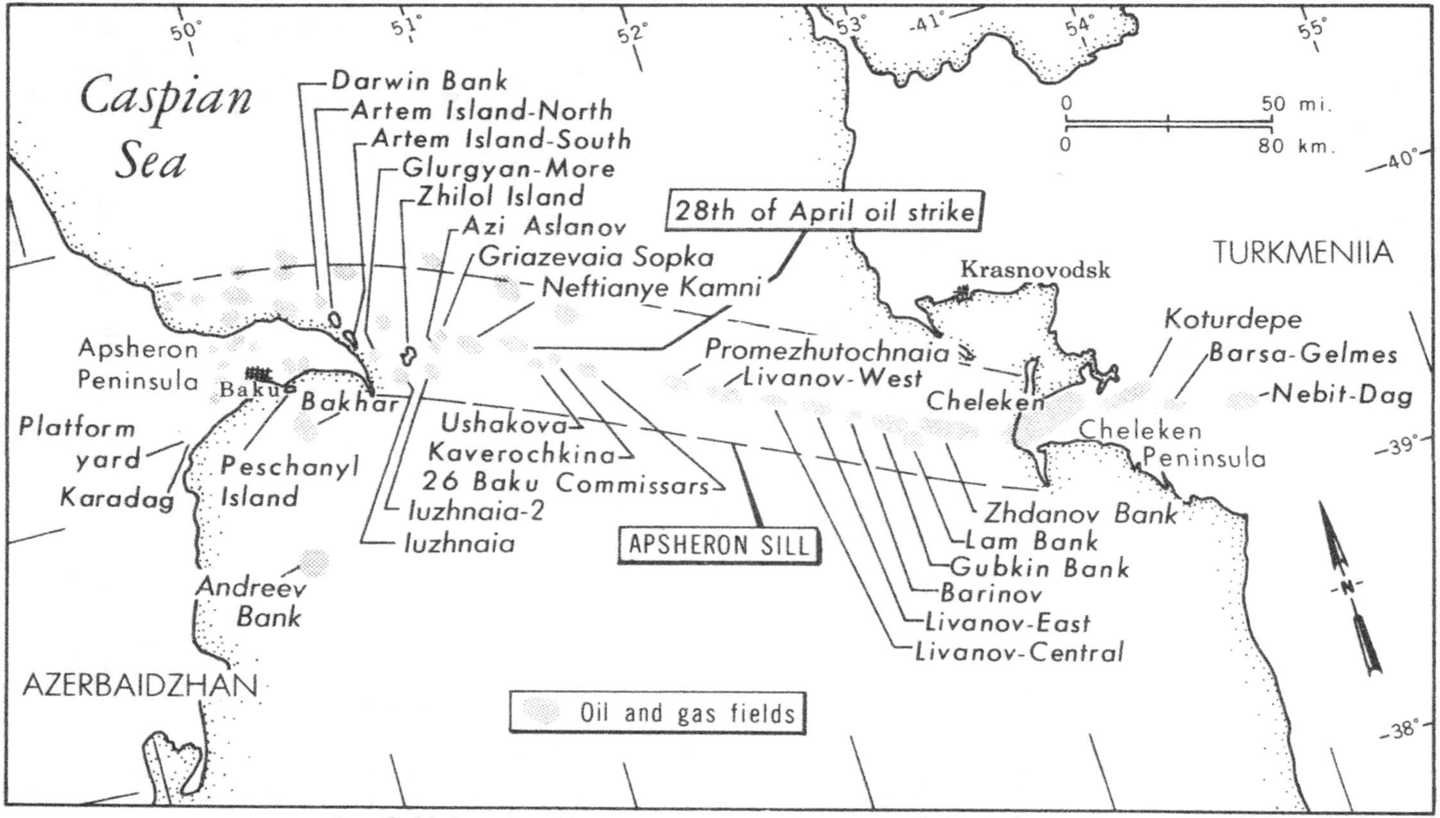

8. Offshore oil structures of the Apsheron Sill

exploration. In 1983, oil was discovered on the northeast flank in water depths of over 100 meters.[24] The find was described as prolific and led to a realization that the 28th of April field could be larger than was previously thought. By the end of 1983, the field had a total of three platforms operating in over 90 meters of water.[25] In late 1985, a fourth fixed platform was added to 28th of April on a well yielding 300 tons per day of petroleum.[26] By 1986, 28th of April had over 21 wells producing 90 tbd.[27]

The most recent discovery in this region of the Caspian Sea is a field about 60 kilometers southeast of Baku. The field known as the "26 Baku Commissars" is believed to be larger than both 28th of April and Kaverochkin, a field discovered in 1985. With an initial test-pay of 2,373 bd, it is believed by Soviet geologists that 26 Baku Commissars could become the most prolific offshore field in the Caspian Sea.[28]

Exploration Along the Kazakhstan and Turkmeniia Coasts

Exploratory drilling and limited production are also taking place off the Kazakhstan and Turkmeniia coasts of the Caspian Sea. Oil was struck in 1966 just off the Cheleken peninsula and at Zhdanov bank about 19 kilometers west of Cheleken peninsula.[29] Since that time three wells yielding test-pays have been discovered at Livanov East bank, 80 kilometers off the Cheleken peninsula. Shortly thereafter, petroleum was discovered at LAM bank, Gubkin bank and Barinov bank, all situated southwest of Zhdanov bank. Further to the south, seismic work was undertaken at Zapadno-Okarern, yet little evidence exists that anything was ever found there.[30]

Another area of exploration in the late 1970s was on the Kazakhstan coast. Here the Soviet Union had high hopes, particularly in the region off Mangyshlak peninsula and Buzachi peninsula. Petroleum was discovered on the Mangyshlak peninsula at Zhetybai and Uzen, and it was assumed oil must lie off the Kazakhstan coast. Similarly, the large fields found in 1974 on the Buzachi peninsula led to the same conclusion. Yet, while a considerable amount of seismic work was undertaken in this region, especially at Adrellamore just off the Tiubedzhik field, little seems to have come of it. The only region off the Kazakhstan coast which still holds promise is Rakushechnoe-more. To the south, Zhdanov bank is the only field expected to yield commercial quantities of petroleum. Exploratory drilling completed in mid-1984 demonstrated the two fields were prospective for oil and gas.[31]

Since 1985, however, most of the new discoveries of petroleum have been off the west coast of the Caspian. Neftianye Kamni's production was recorded as steady. It has been noted a few important wells have been revived, such as well 2123 on the southwest flank which has increased its output to 40 tons per day. Six other wells were discovered on the northeast flank in 1983, and have flowed consistently.[32]

Soviet efforts have been concentrated also off the southeast coast of Azerbaidzhan at Bulla island in the Apsheron bay and at the Bakhar field. Bulla island revealed the presence of large hydrocarbon reserves in 1985 when well 53, one of its exploratory wells, yielded 250,000 cubic meters a day of gas with an expectation of petroleum below 6,000 meters, [33] while the fields in the Apsheron bay were reported as having an above-plan production of 380 tons per day in mid-1985.[34] The Bakhar field proved to be productive enough in 1984 in petroleum and condensate that a single pipe system was built--laid by a Dutch-built, pipe-laying barge--to transport hydrocarbons onshore.[35] Further to the south of this, Andreev bank yielded a test-pay, but little detail is available.

Rethinking the Offshore Industrial Strategy

Despite these developments, however, offshore production has still not reached its 1970 level of 258 tbd. Work at the 28th of April field, for example, has not been as successful as hoped. Exploratory drilling on the field was only 61.2% of the planned target in 1985, causing an estimated loss of 60,000 meters per well.[36] The other fields have been nothing more than test-pay wells with some hope for future development.

While some would say this is a disappointment, it is in fact exactly the role offshore petroleum should be playing at this time. If looked at in the context of upgrading exploratory drilling and enhancing deep-water drilling to find exploitable reserves, the offshore industry has done remarkably well since 1975. Although the initial concern of Soviet planners may have been to raise the output level of the offshore fields in order to buoy up Azerbaidzhan's lagging petroleum output, the result was a more calculated and systematic effort at exploratory drilling and offshore planning. As a result, the finds made off the Kazakhstan and Turkmeniia coasts and the enhanced drilling along the middle portion of the Apsheron sill at Neftianye Kamni and the deeper waters southwest are all examples of a successful reorganization of the industry.

About 1975 the petroleum planners realized in order to increase production and the amount of exploitable reserves it would be necessary to develop a deep-water technology or acquire it from foreign companies. The administrative reorganization that came about in 1978 surely dates from this time as well. Like most other forms of planning in the Soviet energy sector, the changes were comprehensive and dramatic.

Present Technological Developments in the Caspian Sea

The major factor inhibiting the further development of the offshore industry in the Caspian Sea, and consequently forcing the Soviet Union to take a more realistic and sober view of the offshore process, is the absence of an adequate deep-water technology. This is the case both in exploratory drilling and in production from deep-water reserves.

A moderate depth for drilling, assuming no adverse factors such as ice floes present, is up to 25 meters. Here fixed platforms can be placed in the form of large metal trestles. This was the major type of platform that was found in the Neftianye Kamni field until quite recently. After 25 meters, offshore rig design becomes problematic. In waters up to 60 meters, a rig called a jack-up must be used. This is a long-legged structure standing on the sea bottom on either three or four legs. The advantage of the jack-up over the fixed platform is that it can moved by retracting the lower portion of the legs and towing it to a new site. The fixed platform is actually sunk into place and cannot be moved.

The real problems, however, are with rigs known as semisubmersibles. These are designed to drill in waters exceeding 60 meters and, depending on the sophistication of the design, can go to 300 or even 500 meters. The semisubmersible rig contains about 25,000 parts, including the platform itself, which weighs anywhere between 15 thousand and 20 thousand tons and floats in the water on large pontoons. The degree of sophistication in these rigs is quite complex. Special computer-aided propellers must be installed to compensate for wave action on the rig. Further, all the capping facilities must be on the rig in order to maximize output potential.

Pipe laying in deep waters to pump petroleum onshore presents another problem. Special barges are necessary to lay the pipe in precisely the right places, while special tubing material must be used to resist the erosion of potentially high sulphur content hydrocarbons. Piping is an infrastructural concern crucial to the development of offshore fields, especially if those fields are to be turned on and off to supplement onshore production. The important thing to remember in this discussion is, therefore, the deeper and further from shore the rigs go, the more problematic offshore development becomes.

The Magnitude of the Deep-Water Drilling Problem

By the mid-1980s, the Caspian Sea had 176 producing fields. Of these fields, 127 produced in commercial quantities. The majority, 114, were located in waters of less than 60 meters.[37] Most of these are located along the Apsheron Sill at Neftianye Kamni and some of the fields closer to the Kazakhstan coast. Several are also located in the 28th of April field. The remaining 62 productive fields are located in waters of more than 60 meters, with 37 of these fields located at more than 200 meters. Up until now it has been possible to carry on exploratory drilling (*poiskovi-burenyi*) in waters of up to only 200 meters, while production work is currently limited to 150 meters.[38]

Quite recently, the Andreev bank has shown promise, but at wells of more than 200 meters and at depths of over 5,000 meters below the sea bed. The indication is that the southern portion of the Caspian Sea is quite a large oil province, but only offshore and at increasingly deeper rock strata. The task, therefore, is to develop rigs and infrastructural support to service the deep-water deposits.

Essentially, there were two stages of technological development. The first stage concerned the construction of jack-up rigs known as the Kaspii class. These had a limited amount of Western input. While that input was not insignificant, it was not of the order of magnitude of Western assistance used in other forms of offshore technology. The first stage also included purchasing pipe-laying equipment to transport hydrocarbons onshore.

The second stage concerned the development of semisubmersible rigs. Large amounts of Western cooperation provided much of the infrastructural support, such as equipping shipyards for offshore rig construction. In the early to mid-1980s the Shelf class of semisubmersible rigs appeared. These are Soviet produced, in that they are not assembled abroad. Nonetheless there is a dependence on the West in their construction in that all the technology and design is western. This technology and design will soon require upgrading.

The Development of Soviet Jack-Ups

The first really concentrated effort by the Soviet Union to produce rigs other than those of a fixed platform construction was in 1975 when an assembly plant was constructed at Astrakhan. The assembly plant was named the Krasnie Barrikady. The intent was to construct jack-ups to operate in waters of around 60 meters.[39] Prior to 1975 three jack-ups were constructed: the *Apsheron*, the *Khazar* (from the Netherlands), and the *Azerbaidzhana*. These can only operate at very shallow wells in no more than 35 meters of water. They are reported to be still operating by one source, probably in Neftianye Kamni and in the shallow regions of 28th of April.[40]

The first of the modern jack-ups was the *Baku*. It was in large part a copy of the *Khazar*, a jack-up the Soviet Union imported from IHC of the Netherlands in 1967.[41] A Soviet source stated the drilling equipment was built by the Ural Heavy Equipment Plant at Sverdlovsk. The likelihood of the equipment being original Soviet design, however, is dubious.[42] The *Baku* was released to the Iuzhnaia-2 deposit, and by September 1978, it had drilled 3,300 meters of wells.

The rig was described by a Western expert as being of good quality and, in fact, could operate in waters of up to 75 meters.[43] The success of the *Baku* led to the development of two more jack-ups--*60 let Oktiabria* (60th Anniversary of the October Revolution); and, *60 let Azerbaidzhana* (60th Anniversary of the Azerbaidzhan SSR).

The *60 let Oktiabria* was finished in November of 1977; its legs are 108 meters long. It was also built at the Astrakhan yard with assistance from Uralmash. The success of this rig has been impressive. It was towed to Andreev bank in December 1977 and drilled an average of 1,000 meters a day, reaching 5,240 meters by May 1978. It is reported to be still operating in the Andreev field. *60 let Azerbaidzhana* started operations in 1981, [44] but was reported by Soviet press in April 1987 to have sunk several years ago after striking a mud volcano.[45]

Two more jack-ups were built at the Astrakhan yard subsequent to this--the *Sivash* in 1981 and the *40 let Pobedy* (40th Anniversary of the Victory in World War II) in 1985. The *Sivash* is reported to be operating in the Black Sea while the *40 let Pobedy* is operating somewhere in the Caspian.[46]

Most of this development was accomplished largely by the Soviet Union. This should not obscure the fact that there were crucial Western imports at this time. Besides the import of the Dutch-built jack-up, the *Khazar*, in 1967, the Soviet Union turned to the West quite early in the development process. In 1971 a seagoing, pipe-laying barge was bought from IHC of the Netherlands. Named the *Suleiman Yeyirov*, it was designed to lay pipe from Bulla island to Baku onshore. This was part of a plan to integrate piping of petroleum with the exploration going on around the southwest deposits of the Caspian. In total, 180 kilometers were to be laid, but little has been mentioned of this since it was announced in 1975. As of 1982, it is reported petroleum was being moved to the Azerbaidzhan coast, but no mention was made of the amount of piping in operation.[47]

In 1975, the French firm Serete sold a floating drilling platform to the Soviet Union, and in 1976 Blohm and Voss AG (FRG) supplied self-propelled cranes for positioning offshore rigs. All of this was part of the Soviet Union's concerted effort to increase its ability to develop its offshore industry. The French and West German purchases were efforts to enhance the jack-up operations; but it soon became important to develop semisubmersibles. Here Western assistance was critical.

Western assistance for the jack-ups was primarily to supply parts unattainable in the Soviet Union. Key Western involvement in the Soviet offshore industry occurred with the development of semisubmersible rigs and the preparation necessary to build them. It is at this point a type of dependence began to develop in the industry. Semisubmersibles, crucial to offshore deep-water development, are simply too difficult to develop alone. Further, it was much cheaper to import Western technology than to initiate a Soviet domestic campaign.

The Development of Semisubmersible Rigs and Western Involvement

The first move was to prepare for the building of the rig itself. Hence, in 1976, Rauma-Repola Oy (Finland) provided the hull for a semisubmersible. The commitment of Western technology was increased, since the first semisubmersible, the *Kaspmorneft'*, was not released until 1979. In 1978, the Soviet Union contracted Armco (U.S.) to produce most of the onboard equipment. This arrangement was made easier because Armco, via their Japan dealings, was involved already with the Soviet Union in the Sea of Okhotsk reserves. At this time the Soviet Union purchased from Geosource (U.S.) two Command II field processing systems for geophysical mapping at depths exceeding 200 meters, presumably to use in deep waters off the Apsheron Sill. Foremost Industries Ltd. (Canada) supplied 24 tracked geophysical survey vehicles to carry the Geosource equipment.[48]

The Astrakhan yard had to be upgraded also to assist in the development of semisubmersible construction. To accomplish this, large equipment purchases were made from Blohm and Voss AG (FRG) in 1979. The yard was completed some time in 1979. Simultaneously, as the Astrakhan yard was being upgraded, work was underway on the first semisubmersible, the *Kaspmorneft*. Work began in December 1977 and was completed in November 1979. All of the equipment was designed by Rauma-Repola Oy (Finland) and

Armco (U.S.). The *Kaspmorneft'* weighs 17,000 tons, has two 70 meter pontoons and can drill to 6,000 meters in 200 meters of water.

The *Kaspmorneft'* was immediately sent to the 28th of April field, where it discovered petroleum on the northeast flank in 1983. The fact that the *Kaspmorneft'* could drill to 6,000 meters in almost 200 meters of water was significant. Soviet reports indicated in 1983 the majority of the hydrocarbon pay zones in the Caspian Sea exist at 6,000 meters and below.[49] The development and the quick deployment of the rig suggest that the Soviet Union was anxious to increase the output from the 28th of April field, and realized Western expertise was essential.

To this end, the Soviet Union began in 1980 to make plans for the construction of another offshore rig plant. The initial action taken on this was to step up their imports of exploration equipment. In 1980 a $118 million contract was awarded to two French firms: Entrepose GTM pour les Travaux Petroliers Maritime (ETPM) and Union Industriel et l'Enterprise (UIE). The contract was for assistance in developing exploration vessels. The long-term intent was to establish a fabrication yard at Baku, actually in the suburb of Primorsk, to produce rigs for use in waters of 200 meters.[50] The work on the plant also involved Sweden and West Germany, as well as the Ye. O. Paton Electric Welding Institute at Kiev. The plant completed sometime between 1980 and 1985 is known as the Baku Plant for Deep Water Foundations (BZGO).[51] The yard was originally designed to outfit four offshore rigs and go into full operation by 1984 [52], but evidence suggests that it was outfitting rigs as early as 1981. The cost of the plant has been estimated at 335 million rubles.[53]

Development of the Shelf Semisubmersible Rig Series

Imports were stepped up from Rauma-Repola Oy (Finland) in early 1980. The content of those orders is not known, but it would be a fair assumption the Soviet Union was stocking up to develop their own semisubmersible series, known as the Shelf semisubmersibles. Some reports claim the Shelf series is predominantly Soviet, in that only the hull of a semisubmersible (possibly the *Vyborg I*) was bought from

Finland in 1983, and the Soviet Union innovated and expanded it into a semisubmersible.[54] This, however, is doubtful for two reasons: first, the equipment ordered from Finland in 1983 was probably for the Shelf series; second, late in 1983 four Pacesetter units rated at 6,000 meters in 200 meters of water were ordered along with what was reported as Shelf semisubmersible rigging from Finland.[55]

The Shelf class series, while not completely a Soviet creation, is nonetheless a credit to the Soviet Union's management of the offshore technology it has imported. The ability to construct, with little previous experience, albeit with foreign parts, a semisubmersible series is something of an anomaly for Soviet industry, given its low level of innovative capability. The first Shelf class rig was completed at the Astrakhan yard and was outfitted at the Baku yard (suggesting that the BZGO was finally operating in 1981). After this it was likely taken to the 28th of April Field.[56]

Shelf 2 was completed in 1983 at the Astrakhan yard and was also outfitted at the Baku yard.[57] It was taken to the Mikhail Kaverochkin well located in the 28th of April field, and is reported to have made a significant find in July 1985 in 150 meters of water with the well flowing 350 tons per day. This rig alone increased production from the 28th of April field to 65 tbd.[58]

The final Shelf series rig, *Shelf 3*, to operate on the Caspian Sea was probably built in three locations: the Astrakhan yard, the Baku yard, and the Vyborg yard. The Vyborg yard was completed in 1983 and was designed to build offshore rigs for Arctic waters. It is further believed the other four Shelf series now operating in other waters were also built at this yard.[59]

Late in 1984, another offshore rig plant to be built in Narimanov at the mouth of the Volga, 31 kilometers north of Astrakhan, was announced.[60] The plant may be either an addition to the Astrakhan yard, which was not designed for the construction of semisubmersibles, or an ultimate replacement for Astrakhan in order to accommodate more Shelf-type rig construction. Whatever the reason, the indication is that planning for the offshore industry has not ceased. In mid-1986 Soviet officials announced that 30 to 35 mobile rigs are to be added to the 11 mobile rigs now operating in the Caspian Sea.[61] Presumably these are to be built at the new yards at Narimanov and at Vyborg.

The Shelf series, operating first in the Caspian Sea and now in other waters, is of some significance in the development of the Soviet offshore industry. First, it was a successful attempt by the Soviet Union to construct its own semisubmersible rig series. Surely many of the components did come from the West, but the assembly was domestic. Second, the Shelf series shows the Soviet Union is serious about offshore development. The price of the Baku yard, for example, at 335 million rubles and the high costs of importing equipment for the yards and rigs exemplify the extent to which the Soviet Union is willing to go in the industry's development. Finally, the commitment to Western imports and the concern with deeper oil deposits suggests the Soviet Union realizes the potential of its offshore reserves below 6,000 meters and in water depths exceeding 200 meters.

Lessons from the Caspian Sea Experience

This region is very significant by virtue of the fact it is the only substantial offshore producing region in the Soviet Union. The Caspian Sea region has the longest history of development and the most temperate conditions, providing an especially favorable laboratory for learning about the offshore industry. The Soviet Union's experience in the Caspian Sea is important because it has determined the whole strategy of offshore development in the Soviet Union. Three main points can be derived from the Soviet Union's experience here.

First, it was discovered around 1975 that hope of raising the level of production from the offshore regions to buoy up the lagging onshore production in Azerbaidzhan was futile. Instead, what became clear was the magnitude of the problem of developing the offshore fields to simply prepare them for production.

Second, the Caspian Sea experience forced the Soviet Union out of its autarky in the petroleum sector, albeit in a small way compared to the expenditures on domestic onshore petroleum equipment. If any of the fields showing promise through seismological surveys were to be developed, it would be necessary to import large amounts of Western equipment. Surely this cannot have been an easy thing for Soviet energy planners to concede. What is important is the almost complete dependence on imports of Western equipment, even on Soviet-assembled rigs. This is a theme which recurs throughout the discussion of the Soviet Union's offshore petroleum industry. It becomes especially pronounced when cold weather and excessive ice floes are added to the deep-water drilling problem.

Finally, due to technological constraints, the Caspian Sea experience forced the Soviet Union to turn from an emphasis on petroleum production to one of emphasizing exploration. In the petroleum industry in the mid-1970s, this was unusual. The amount of exploratory drilling, as noted in Chapter 2, was very low overall in the petroleum sector. The Ministry of Petroleum's main task in the mid-1970s was to meet the draconian output targets set each year by Gosplan. By emphasizing exploratory drilling and a more calculated approach to development planning, the offshore petroleum industry became one of the more dynamic of the Soviet resource sectors. In 1982, for example, offshore petroleum was one of the earliest industries allowed to form joint ventures with foreign firms in the form of temporary permitted licenses. This was a full five years before Gorbachev's wide-ranging, joint venture proposal in January 1987.[62]

It is necessary now to turn to the other regions of exploration and production. First, the Black Sea and the Sea of Azov will be discussed. The Pacific Far East will then be dealt with. This region is quite important because it was the only other region besides the Caspian Sea with some offshore petroleum production experience (in the late-1970s). It was also the next region where the Soviet Union relied on the West for development assistance.

The Black Sea and the Sea of Azov

Soviet offshore activity in the Black Sea and the Sea of Azov has been quite low compared to the Caspian Sea, the Sea of Okhotsk, and the Soviet Arctic. The main reason for this has been the predominance of gas reserves in these areas and the abundance of onshore gas for the foreseeable future. Most of the petroleum discoveries have been in the Rumanian sector of the Black Sea. Rumania discovered oil

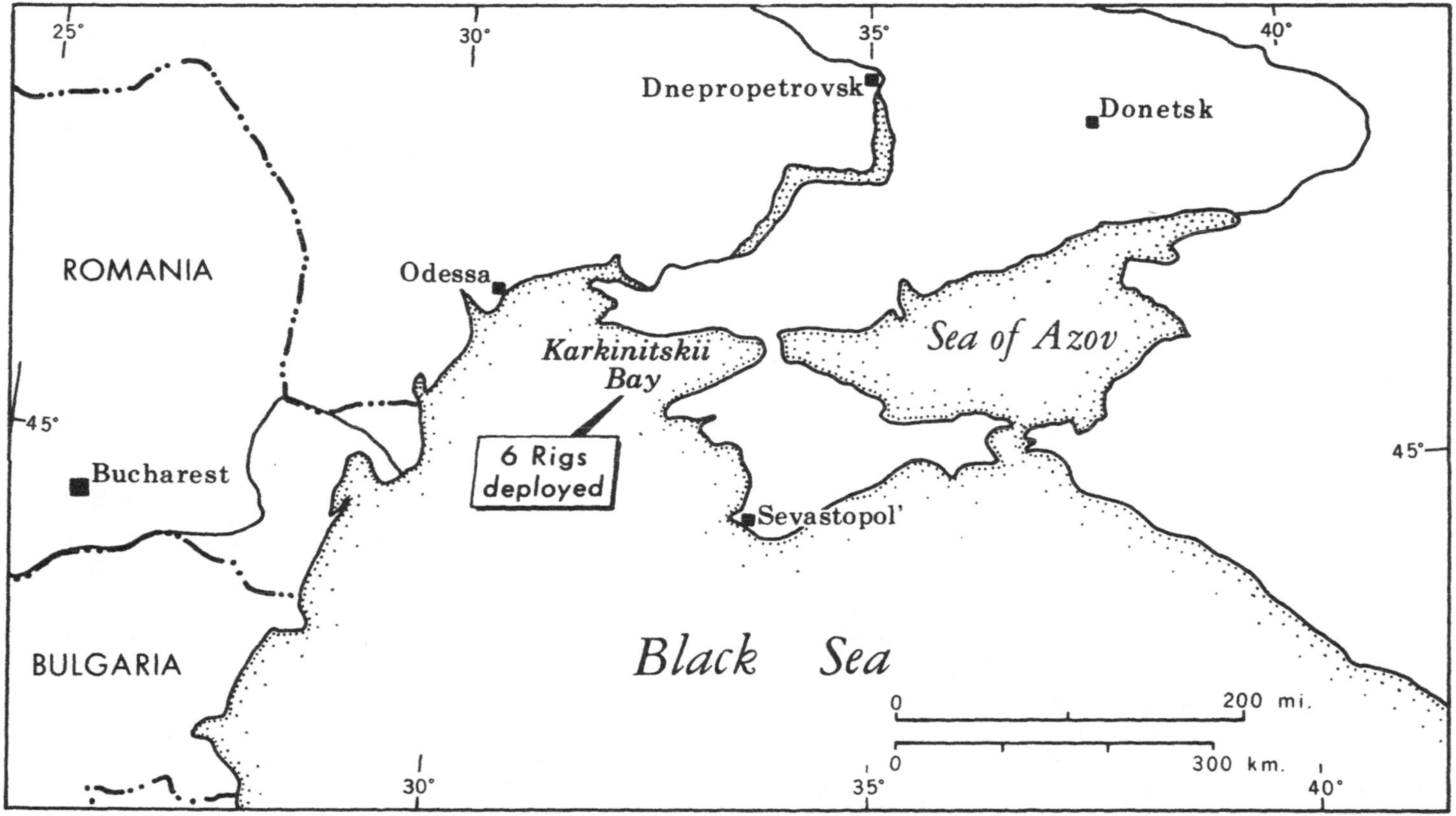

9. The Black Sea and the Sea of Azov

there in November 1979 after drilling for only three months.[63] The Soviet sector of the Black Sea, however, has only yielded gas. The Sea of Azov (an arm of the Black Sea) has yielded one gas pay. Although this area may be a substantial gas producer in the future, its petroleum prospects are not seen as optimistic.

The discovery of gas in March 1975 at the Golitsyn Uplift led to development plans for the area. As a result, some fixed platforms were built in the Golitsyn area. Construction was slow, and it is doubtful these platforms were completed.[64] The Soviet Union also prepared two areas for drilling off the Georgian coast in ice-free waters.[65] In 1979, the Soviet Union did focus some attention on the Sea of Azov, largely in hopes of finding a clean energy source (i.e., gas) to service the resorts of the Black Sea. Commercial gas reserves were found in the Sea of Azov near the Crimean peninsula's Arbat Spit and at the east end in Krasnodar Territory's Beisugskii estuary. Small-scale gas production has started there.[66] The Sea of Azov, while not terribly important from a petroleum point of view, did offer the Soviet Union a place to test ice-resistant rigging before the Vyborg yard came on-stream. Rigs had already been deployed in the Barents Sea. Judging by the amount of expertise and experience required from abroad, however, the Soviet Union cannot have learned a great amount from these facilities. With the Vyborg yard producing semisubmersibles and accepting jack-ups for outfitting in the Arctic, the facilities in the Sea of Azov can expect to be scaled down.

The Soviet Union has six mobile rigs operating in the Black and Azov seas, the jack-up *Sivash* deployed in 1981 and the five semisubmersibles of the Golitsyn class deployed in 1986.[67] All of these are looking for gas, reportedly with little success. Ever since 1979 when the Soviet Union began to rethink its offshore strategy and focus instead on exploration and production in the Caspian, Okhotsk and Arctic waters, the Black and Azov seas have always been given low priority. With a predominance of gas, the Soviet Union cannot spare rare mobile rigs of advanced technology to these regions. The five Golitsyn class semisubmersibles deployed in 1986 are, in my opinion, being tested in this region and will probably appear in the Arctic waters or Caspian Sea in the next few years.

The Pacific Far East

The most important region of the Pacific Far East is the area off Sakhalin island in the Sea of Okhotsk. This is the only other region that produced any petroleum from its offshore waters. The waters off of the Kamchatka peninsula and in the Sea of Chukotsk have not been explored at the time of this writing and, except for a few wells drilled on the Chukotsk peninsula and the eight wells (mainly gas) on the east coast of Kamchatka, planning is still far off.[68] Discussion will therefore center on the Sakhalin area. Essentially, there were three stages of development in the region.

First, was the discovery of petroleum on the northeast shelf of Sakhalin in the Sakhalin gulf, in the Tatarskii strait, and the southwest portion of Sakhalin. Most of the exploratory work has centered around the Odoptu field off Okha, the Chaivo Bay and Katangli. The reserves are evidently large, which has encouraged the Soviet Union to develop the region. In doing so the Soviet Union drew upon its experience in the Caspian Sea. Rather than waste venture capital on developing a region known to be beyond its technical control, foreign firms were enlisted.

The second stage of development in the Sea of Okhotsk, therefore, was the involvement of Western partners, specifically the Japanese. The involvement with Japan is one of the longest standing offshore agreements the Soviet Union has in that industry. It has revealed quite a bit about the operation of joint cooperation with a capitalist power. Much of the involvement with Japan means the indirect participation of the United States is important given the amount of offshore equipment licensed to Japanese rigs from American firms. The Soviet Union tried throughout the life of its cooperation with Japan to reduce this dependence upon American technology. In this bid the Soviet Union was unsuccessful and soon became aware how widespread American technology, via licenses, is throughout the world. This forced the Soviet Union to take a more cautious approach in other regions.

Finally, the third stage of development provided one of the learning grounds for Arctic offshore development. Despite the problems inherent in East-West cooperation, the

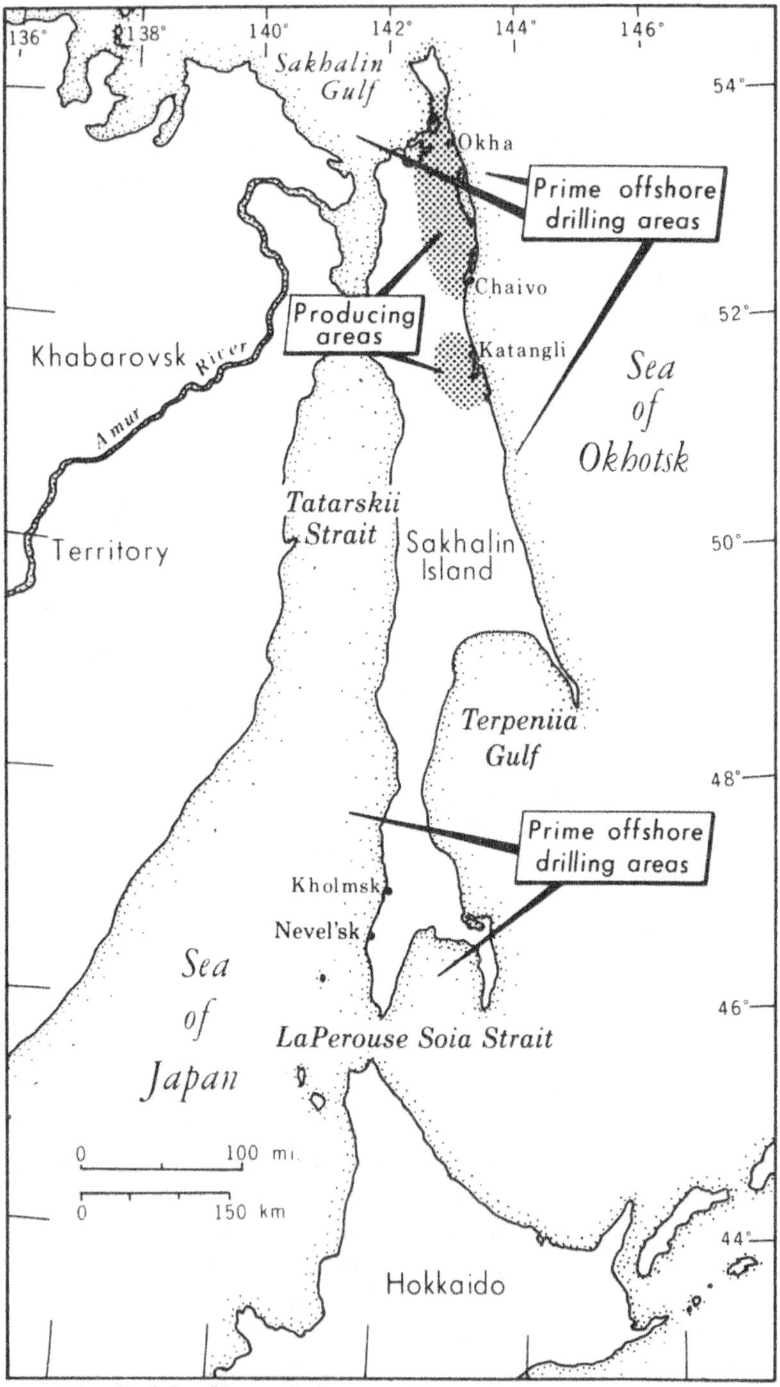

10. Sakhalin offshore oil and gas resources

experience in the Far East in the Sea of Okhotsk, with its large ice floes and multifolded seabed, forced the Soviet Union, as early as the mid-1970s, to test and buy equipment far more complex than any required in the Caspian Sea. The one difference is that the shelf around Sakhalin is shallow by comparison to the shelf of the Caspian Sea off the Apsheron Sill; hence, semisubmersible rigs have only recently been introduced.

Discovery of Petroleum: Background

Commercial production on Sakhalin island began in 1921, when the island was under Japanese occupation. Production continued through World War II, rising to some 0.8 mty in 1945.[69] After the war, petroleum production increased steadily until 1964 when it reached about 2.4 mty. By 1967, five fields onshore had been discovered with 28 oil wells and wildcatting underway in 51 new areas. Production peaked between 1965 and 1970 at around 2.5 mty. In 1970, however, production onshore was showing signs of strain. Only three new fields were discovered, and only 10 of 38 new wells yielded petroleum.[70]

Around 1970, the Soviet Union began to negotiate with Japan the joint development of the Sakhalin waters. This was not a particularly unusual event. The Soviet Union had been exporting crude oil to Japan since 1960, when 1.1 million tons of petroleum were sent, representing nearly 4% of total Japanese petroleum imports that year. Japan was also involved in negotiations with the Soviet Union over Tiumen' oil developments and Iakutsk natural gas.[71] Given the fact production was declining onshore and slant drilling offshore revealed large reserves probably existed offshore, both sides began to posture for an agreement.

The Japanese-Soviet Cooperation over Sakhalin

The Japanese-Soviet cooperation over Sakhalin is the most dominating event of the development in the offshore industry of the Far East. The contents of the agreement were crucial to a successful venture from the Soviet Union's point of view. In the first place it was not prepared to go offshore alone. It saw in Japan a partner with a traditional interest in the region and a country concerned over security of oil supplies. Furthermore, Japan had access to the most up-to-date technology via the United States and an agreement was surely attractive. In addition, the Soviet Union probably foresaw difficulties with the Japanese if the Soviet Union did not include them in the development of the region. Sakhalin and the Kuril islands were still under territorial dispute. Any joint development effort might force Japan to grant *de facto* recognition of the sea and island borders.

The timing of the negotiations between Japan and the Soviet Union are significant in two ways--the Soviet Union's domestic production constraints brought on by its aging technology, and Japan's energy situation on the eve of the OPEC oil crisis.

1970 was a benchmark year for the Soviet Union. In two regions, Azerbaidzhan and onshore Sakhalin, production of petroleum began to fall. It was important for the Soviet Union to halt the decline in these areas. In the Sakhalin, as in Azerbaidzhan, the answer seemed to be in the offshore reserves. In the Far East, however, the Soviet Union had one very important constraint: the excessive ice floes in the Sea of Okhotsk from mid-September until mid-May. This means that the time available to develop resources was shortened from 12 months, as in the Caspian Sea, to 4 months in these waters. As a result, it was in the Soviet Union's interest to negotiate a deal with Japan, which had access to proven offshore technology. With such a short drilling season, there is no time to drill the wrong well, especially if the intent is to boost oil output quickly to benefit from high world market prices. Moreover, Japan was facing its age-old fear of lacking adequate energy reserves. With OPEC threatening to drive oil prices up, Japan was more than interested in entering into an agreement where it would receive petroleum, expectedly at lower than world market prices, and share in the long-term rights to the deposits. In fact, it was at this time that Japan began its global search for new energy supplies in anticipation of an aggressive OPEC cartel. The proximity of the reserves was therefore attractive because transportation margins would be relatively low for Japan.

Soviet Negotiations with Japan Begin

The negotiations began officially in October 1970 when H. Imazato, president of Japan's North Slope Oil Co., visited Moscow to follow-up on several visits between Soviet and Japanese energy officials concerning Sakhalin development that had taken place earlier that year. The Japanese were quite concerned over the fact a Soviet-West German trade accord signed earlier in 1970 had made West Germany the Soviet Union's number one trading partner, thereby surpassing Japan.[72]

In September 1971, the five-year trade accord between the Soviet Union and Japan was due to expire. At the last meeting of the trade round, the Japanese brought up the subject of joint exploration of the waters off Sakhalin; it soon became a topic of discussion between the two.

The Soviet Union made the first proposal. It was predictably as isolationist as possible. Japan was to provide the Soviet Union with credits, estimated at around $200 million, to buy equipment, machinery and exploration materials. Japan was to provide drilling vessels and bear half the costs. The Soviet Union was to undertake all exploration exclusive of Japanese personnel. Japan for its part would receive oil and gas as payment. The remaining issue then concerned Japan's remuneration should no hydrocarbons be found. The Soviet Union proposed returning the equipment (less depreciation) and taking on half the exploration costs.[73]

Japan was not satisfied with this arms-length proposal. Japanese participation was essential. The Japanese counter proposal at this time was to offer to participate in exploration and production, to put up $80 million in credits, and to bear half the costs of exploration and development. Japan would take some volume of the output as a bonus.[74]

The negotiations then, hinged on the issue of credit. Clearly, it was unacceptable for the Japanese to have no plan for Soviet repayment of credits. But the Soviet Union was prepared to pay only for equipment use if there were no hydrocarbons found.[75] The issue for the two countries was, therefore, over credit and risk. That is, the Soviet Union wanted to take all the credit possible and incur no risk as to its use. In spite of this, by early 1974 Japan had decided it would invest $100 million (up from the earlier proposed $80 million). The terms of repayment were not clear, however, and it seemed as though the Soviet Union was willing to wait for a better negotiating position.[76]

There was a dynamic underlying the negotiations at this time which did not allow either party to propose an all-or-nothing deal. For Japan, the issue was competition. In January 1974, Atlantic Richfield Co. (U.S.) expressed interest in cooperating with the Soviet Union to develop the Sakhalin offshore deposits. Atlantic Richfield's chief executive officer, Anderson, met with the Soviet Foreign Trade Minister Nikolai Patolichev in Moscow.[77] The details of the meeting were never revealed, but it was probably not a productive one. Atlantic Richfield could not have been able to put up anywhere near the capital the Japanese government could. The meeting was probably to negotiate the sale of some equipment and parts for offshore vessels. Japan probably interpreted this as the beginning of a Soviet effort to enlist as many American companies as possible, each on an individual basis, thereby giving the Soviet Union access to the necessary technology while maintaining complete independence in the region. This was probably not the Soviet Union's intention. It was important for the Soviet Union to get the development of the region underway for reasons of regional domestic demand, and to take advantage of the increasing world market price of oil being driven up by OPEC. With inadequate capital resources, it was in the Soviet Union's interest to continue negotiations and not to take too hard a stance.

As a result, after several meetings through early 1974, Soviet Foreign Trade Minister Shushkov and H. Imazato met in Tokyo on April 29 and signed a memo guaranteeing that in the event of an agreement half of any crude petroleum discovered in the joint development project would be delivered to Japan for 18 years. Further, the initial $100 million credit, it was agreed, would be repaid by the Soviet Union over five years. If work was not successful, Japan would receive only half of its investment back. Additional work after the five-year period would bring another $100 million in credits from Japan, with the Soviet Union bearing half the production costs in this second phase. Finally, Japan would receive an 8.4% discount on oil and gas prices for 10 years after production commenced.[78]

The negotiations were not over at this point. One of the most important issues was still unresolved--American involvement. It was clear both sides wanted and required American involvement. It was, therefore, no surprise the Soviet Union agreed to American participation in the project when the Japanese requested it.[79] This was a wise concession on the Soviet Union's part. Since they were buying Japanese technological expertise, it would be best to allow the Japanese control of their own technological inputs, thereby gaining access to the most up-to-date American technology.

The Formation of SODECO and the Involvement of Gulf Oil Co.

Shortly after the Soviet Union agreed to American involvement in the project via Japan, H. Imazato announced Gulf Oil Co. (U.S.) would provide technology and capital for a new company to be established by several Japanese firms. The participation by Gulf Oil Co. was seen by many Western firms as a wise move. British Petroleum (BP) (U.K.) actually had wanted to preempt the Gulf Oil Co. in the deal.[80] This provided the impetus to move ahead with the consortium of Japanese firms.

The first Japanese companies involved in the venture were: Overseas Petroleum Development Corp.; Marubeni Corp.; C. Itoh and Co.; and Japan Petroleum Development Corp. (the state-owned petroleum company). SODECO was to eventually include 17 companies. Its initial capital was $134.7 thousand, with an ultimate capital target of $30.3 million.[81] The involvement of Gulf is interesting because it was to contribute ultimately 10% of the equity of the corporation once it grew in asset size to $30.3 million.[82] Eventually, Gulf invested, in December 1975, $211 thousand in the consortium to acquire a 3% share in the operation. By 1975 before an agreement had been signed, SODECO's assets had already grown to over $7 million. In addition, Gulf agreed to commit one of its advanced seismic vessels for exploratory work.[83]

This was quite a commitment for Gulf and a substantial gain for the Soviet Union. Therefore, neither Gulf nor the Soviet Union were going to take any chances by solely relying on the Japanese to coordinate the interaction of the American company and the Soviet government. American participation was too crucial an item to leave to an agreement that had not yet even been finalized.

In early October 1974, Gulf officials met with Soviet officials in Moscow and signed a general technical cooperation agreement independent of the SODECO deal. This was probably an affront to the Japanese who later got a concession from both Gulf and the Soviet Union that SODECO would take a 10% interest in any results from the separate Gulf-Soviet agreement.[84]

This was an example of what the Soviet Union would do throughout the cooperation. It was in their interest to involve as many firms as possible at this point in order to gain technical expertise. The Soviet Union was knowingly playing a balancing act. On the one hand, the Japanese government could decide unilaterally to end the deal (although unlikely, it had to have been a Soviet fear). On the other hand, working with American companies had a potential for problems with U.S. export controls, as was later the case. The idea, therefore, was to spread the risk as much as possible without angering their Japanese partners.

Negotiations between the Japanese and the Soviet Union carried on throughout 1974 until, in October, the two sides sat down to work out a final agreement. In addition to the previous provisions, a $30 million credit line was offered to the Soviet Union to buy Japanese consumer goods. In addition, $20 to $25 million in credits were extended to the Soviet Union to buy special equipment, such as computer systems for exploration vessels.[85] The agreement stalled briefly in late 1974 (around November) when the Soviet Union tried to get Japanese credits of $252.5 million over ten years regardless of risks.[86] This, however, was not a major stumbling block and was in fact part of the final agreement.

The Sakhalin Development Agreement Concluded

On January 28, 1975 an agreement was reached. Besides the $100 million dollars from the Japanese private and government sources (the private companies in SODECO were to contribute 30% of this) [87], the Japan Export-Import Bank was to provide between $600 and $700 million, representing approximately half the project's requirements. The $100 million credit now had a 6% interest rate attached to it. The Japanese allowed the inclusion of a "no success--no repayment" clause, whereby Japan would not be entitled to any repayment if exploratory work revealed no reserves. Japan evidently had faith in the project. After the Japanese received 50% of the output at a reduced rate of 8.4% over 10 years (as repayment for its loan), Japan would then be entitled to continue to receive 50% of the output for another 10 years, but at the prevailing world market price. The volume provisions of oil and gas deliveries were to be decided annually throughout the life of the agreement with Soiuzgazexport, the Soviet gas export agency, and Soiuznefteeksport, the Soviet oil export agency, determining the overall production rates.[88]

It was hoped the consortium would undertake exploration immediately. Several Soviet foreign trade organizations became involved in the buying of equipment, including Sudoimport, Elektronorgtekhnika, Promsyrioimport, Mashpriborintorg and Mashinoimport. The Soviet Union set its goal for the project. In the first five years, 45,000 linear kilometers were to be surveyed and 26 pilot bore holes were to be drilled.[89] Work, however, did not begin until late in 1975. The Soviet Union tried to work into the agreement, after the fact, a proposal it made as far back as September 1971 that it have exclusive personnel rights to the exploratory work in the Soviet region. As in 1971, the Japanese found this unacceptable. Once again, the negotiations stalled. This time, after the agreement had been signed. The problem involved the Soviet Union's wanting to send engineers to Japan for training. Knowing how committed the Japanese were to the project, it seemed as though the Soviet Union was trying to dupe the Japanese. When Japan refused, the Soviet Union compromised. Seven non-Soviets were to be allowed on survey ships.

This post-agreement tussle seems innocuous enough, but it is important. The Soviet Union's interests had never really changed throughout the entire negotiating process. The Soviet Union was unwilling to accept a provision requiring it to repay the $100 million loan if no hydrocarbons were found. On this point the Soviet Union won. The issue of foreign observers was an even more sensitive one. Here the Soviet Union was forced to give in, which could not have been an easy decision for it to make. This is, then, one of the most important implications of the whole process. The Soviet Union was now intrinsically involved with a Western power, i.e., foreigners were physically on-site in a security zone. Despite the Soviet Union's willingness to explore and develop the region, they would, however, never be happy with foreign advisers in the area. It would be one of their long-term policies to limit the foreign presence.

Exploration Begins

Geophysical surveys began in May 1976. At this time two Japanese ships and one from Compagnie Generale Geophysique (CGE) of France were leased. The French ship was named *Orion Arctic*, the two Japanese rigs were known as the *Ranger Maru* and the *Hercules*.[90] The ships made test borings in three areas, two on the northeast shelf and one on the southwest shelf.[91] In anticipation of early finds, the Soviet Union leased from SODECO an offshore drilling platform to drill in 70 meters of water to depths of 6,000 meters. The rig was leased from Interocean (Japan) and was built by Suitomo Heavy Industries (Japan). What is interesting is that 20 specialists were provided by Gulf Oil Co. to equip the rig.[92]

Almost immediately the Soviet Union began trying to reduce its dependence on too few partners. CGE's *Orion Arctic* was leased outside of the agreement with SODECO.[93] The Soviet Union, via Sudoimport, also undertook the purchase of $14 million worth of digital seismographic recorders, magnetometers, gravity meters, cables and hydrophones from other French firms. The French for their part decided to provide training on-board the vessels for the Soviet Union.[94] The Soviet Union then contracted Mitsubishi Corp. (Japan) to install the equipment on-board two Soviet vessels. Further, the Soviet Union used $2.5 million of its credited $150 to $155 million ($100 million initial loan, $30 million in consumer goods credits and $20 to $25 million in equipment credits) to finance the operation.[95]

At the same time the Soviet Union was dealing through SODECO. In March or April of 1977 Sudoimport bought from Kyoho Tsusho Kaisha Ltd. (Japan) four 20-ton floating cranes; Kyoho Tsusho in return accepted Soviet-made hard alloy drills from Stankoimport. Four more floating cranes were bought from Fukoshima Ltd. (Japan) valued at $10 million.[96] A jack-up was ordered at this time from Mitsui Engineering and Shipbuilding Co. Ltd. (Japan). The rig was based on a design by Levingston Shipyard Co. (U.S.), and the equipment on-board supplied by Otis Engineering Corp., (U.S.) and Armco (U.S.). In the summer of 1977, when the rig the Japanese leased from Interocean, known as the *Interocean I*, ran aground on its way to drill in Sakhalin, the Soviet Union leased from Dolphin International (U.S.) the *Borgsten-Dolphin*, a jack-up capable of drilling in 90 meters of water to depths of 8,000 meters.[97]

Quite early in the development process it became evident American technology was crucial to the success of the project. Before any drilling had taken place, five American firms had already pledged equipment, three of them outside the SODECO agreement. This was to become characteristic of the Japanese-Soviet cooperation. In the event of any failure on the part of the Japanese either to obtain or develop essential equipment, the United States was on hand to substitute. In part, although this is what the Soviet Union wanted to achieve, there were a number of dangers, the most serious being the threat of American sanctions.

The Discovery of Oil in the Odoptu Field

Nevertheless, the plethora of activity did pay off. In August 1977 the first exploratory well was spudded in the Odoptu field just north of Okha. The well discovered by the *Borgsten-Dolphin* had a flow of 7,000 bd, with estimated reserves of 100 million metric tons of hydrocarbons (including gas and condensate).[98] By the end of August, SODECO reported that four wells had been drilled.[99]

The discovery of oil in the Odoptu field was important--not only for the obvious reason of the Japanese-Soviet cooperation's probable return on its investment, but from the point of view of how the oil was discovered. All of the crucial equipment, i.e., rigs, seismographic instruments and even pre-drilling gear, such as floating cranes, were from foreign companies. Unlike the Caspian experience, there was no sole Soviet production program using offshore rigs like the Shelf series. With the Odoptu field discovery, the Soviet Union realized it was truly dependent upon Western technology. The Soviet Union then attempted to ensure a consistent inflow of Western technology, while trying to edge out the active participation of the Japanese. The intent was to acquire technology and training from the Japanese and others with a minimum amount of foreign contact. This dual policy was to dominate the life of the agreement.

Increased Foreign Buying and Disputes with Japan

In February 1978, the Soviet Union requested and received an extension of credit to the original agreement with Japan. There were, however, problems with the agreement at this point. The Soviet Union, intent upon pursuing its desire to eventually edge out the Japanese side while retaining foreign equipment, requested the Japanese extend the credit line to allow the Soviet Union to eventually purchase its own rigs (presumably from the Japanese). These were to replace the existing foreign rigs operating in the zone. This request was unacceptable to the Japanese, who instead proposed that the agreement be changed to increase proportionately the price reduction for oil deliveries to Japan with the extension of Japanese credits. The Soviet Union turned this proposal down.[100]

Negotiations over credit extensions went on throughout the year and probably would have been a point of serious contention with the Japanese, especially since two wells off of the southwest shelf of Sakhalin, Staroniachin and Kuznetzov, were found to be dry.[101] Yet the *Hakarya II* jack-up, which had been introduced to the northwest shelf of Sakhalin to drill four wells, reportedly made a second petroleum strike at Odoptu.[102] Immediately following this strike the Soviet Union moved to secure more equipment crucial to the expansion of development and turned to Norway and Finland this time, ordering five supply vessels. Three were supplied by Ulstein Hatlo A/S (Norway), the other two were supplied by Laivatellises Oy (Finland), under contract with Skaugen Offshore Supply Ships (Norway).[103]

The investment once again seemed to be paying off. A third strike in the Odoptu Field was made, yielding a test-pay of 1,570 bd.[104] This probably was a significant factor leading to a revised agreement by the spring of 1979 which allowed the Soviet Union an additional $70 million in credits to the already pledged $100 million.[105] The loan had a 6% interest rate attached to it. What is of some importance, however, is that the Japanese extended the payback period from five to seven years.[106] The credit extension requested by the Soviet Union was based upon the price inflation the Soviets claimed had taken place over the mid-1970s. It may also have been an indication that they anticipated more problems in exploration activities than when the agreement was first negotiated in 1975. Probably the long time period before even test-pays were revealed led to this conclusion.

The new credits allowed the Soviet Union to undertake more foreign buying. In September 1979, Sudoimport bought a $36 million self-lifting drill rig from Mitsui Ocean, (Japan).[107] Sudoimport moved to replace the *Borgsten-Dolphin* which had been damaged while drilling late in 1978.[108] Sudoimport ordered a Class III, three-legged jack-up from Levingston Shipyard Co. (U.S.) which was rated in 90 meters of water and capable of drilling to 8,000 meters. The equipment for the rig was to be supplied by Armco (U.S.) through its National Supply Division. Support equipment was provided by more American firms. N.L. Petroleum Services (U.S.) delivered two blow-out preventer stacks and controls valued at $525 thousand. Further, Baker International (U.S.) delivered testing equipment to the Levingston rig.[109]

Things seemed to be proceeding quite well after the revision of the agreement. All of the recent purchases from American companies had been made via SODECO as per the agreement, and another well had been spudded off the coast near Chaivo. The number one well yielded 1.7 million cubic feet per day of gas, 1,500 barrels a day of condensate, and petroleum was expected at lower depths.[110] Problems began to develop, however, of a much more serious nature than the Soviet Union's financing arrangements with Japan.

American Sanctions and the Breakdown of Cooperation

In 1979, the Soviet Union invaded Afghanistan. This brought with it a series of sanctions from the Carter administration. These sanctions were eased off in the first years of Reagan's administration, but were reimposed during the West Europe-Soviet gas pipeline sanctions. Essentially, all of this spelled the end of the agreement with Japan and the development of the Sakhalin deposits, at least with Japanese assistance, for quite a while. When a falling world price for oil in late 1985 and early 1986 was added to the American pressure, the Soviet Union decided foreign cooperation was not only too politically volatile, but also too costly.

The first sanction involved McDermott Inc. (U.K.). McDermott, through its subsidiary Oceanic Contractors Inc. (U.S.), signed a $1.5 million contract to undertake a feasibility study of permanent fixed platform drilling in the waters around Sakhalin. This agreement was then followed by a $450 million equipment deal. The Carter administration stepped in and declared a sanction on the license of U.S. oil and gas parts to the Soviet Union.[111] This intervention brought a strong appeal from the Japanese to the United States to exclude the Japanese-Soviet cooperation in Sakhalin from the general American embargo of the Soviet Union. The Japanese complained both French and West German trade with the Soviet Union was being increased because of America's blanket sanctions.[112]

The fact of the matter was that American technology was intrinsic to Japan's success in the region. Indeed, the Japanese were aware it was probably the most important attraction for the Soviet Union, far outweighing any other considerations. It came as a real blow, therefore, to both the Soviet Union and Japan when the United States Department of Commerce (USDC), halted equipment sales to the Soviet Union by Schlumberger (France-Netherlands), Halliburton Oil Products Co. (U.S.) and Otis Engineering Co. (U.S.).[113]

Seeing no other option the Japanese increased their lobbying of the USDC and in the spring of 1980 met with some success. In May, all American subcontractors to SODECO were given approval to recommence licensing and sales to the consortium.[114] The reprieve was timely since C. Itoh and Co. (Japan) had started negotiating a technical assistance deal with the Soviet Ministry of Gas to deliver a new type of drilling rig in late 1980.[115] Armco (U.S.) was then given the go-ahead to supply Mitsui Ocean Development Co. (Japan) with equipment because the Japanese firm was building a rig for Sudoimport.[116] The rig, named the *Okha*, was delivered in July 1980 and began drilling off Chaivo, east of Sakhalin island. In addition to this the Soviet Union enlisted Offshore Supply Association (U.K.) to operate two supply vessels in the region.[117]

The suspension of American sanctions quickened the pace of development. The *Okha* soon discovered a well yielding 2,050 bd and 77 million cubic feet of gas a day.[118] It was discovered that the Chaivo field had proven reserves of 630 million barrels of oil, 140.5 bcm of gas, and 142 million barrels of condensate.[119] This brought the number of exploratory wells to 13 in the whole region by 1981, 7 of which had promising oil reserves. The total was to be brought to 18 by year-end with 5 more drilling sites planned.[120] Japan was once again eager to continue the agreement and extended its credit terms, this time to the end of 1983.[121]

The Japanese, however, were not to be allowed to pursue the development without more hindrance from the Americans. In 1982, Reagan extended his Soviet gas pipeline sanctions to all U.S. oil and gas equipment being sold to the Soviet Union by U.S. subsidiaries and overseas licensees of U.S. companies.[122] Japan quite rightly claimed all the foreign equipment remained in foreign hands unlike the gas-pipeline deal.[123] There was little persuading the Americans at this point, however. Once again Japan found itself at a loss. It was unable, for example, to replace the wire line logging equipment for checking the hydrocarbon potential of bore holes that were only produced by General Electric Co. (U.S.) and Schlumberger (via American licenses to the French-Dutch company).[124] The Japanese estimated it would take them at least two to three years to replace this and other types of technology.[125]

The Japanese had, however, dug themselves in and agreed to help the Soviet Union to equip and build its own rigs. When the *Hakarya II* (the rig built by Mitsui Corp. (Japan) and designed by Levingston Shipyard Co. (U.S.), was withdrawn due to the large number of American parts licensed to the rig, the Japanese replaced the rig with one of their own.[126]

The Reagan administration eased early in 1983 the sanctions and allowed the Japanese to buy from previously banned companies in the United States. Once again activity started up. The Soviet Union bought four drilling platforms for the Chaivo field and undertook a separate liquid natural gas project with the Japanese.[127] Further, the Soviet Union took measures to lengthen the season in the region by purchasing from McDermott Inc. (U.K.) preliminary engineering equipment and conceptual designs to allow the Soviet Union to produce its own ice-resistant rigging and equipment.[128] This was a significant purchase despite the eventual down-sizing of the Sakhalin development. It was clear this technology was to be used for exploration and eventually production in the Soviet Arctic.

This purchase by the Soviet Union coincided with the introduction of one of the Soviet built semisubmersibles, the *Shelf 5*, to the region.[129] It appeared the Soviet Union was preparing to go it alone in the area as the agreement was coming to an end. The Japanese were willing to continue the operation after the deposits that had been surveyed were assessed. The Soviet Union, however, was not so enthusiastic. What had occurred since the Reagan administration's imposed sanctions was the Soviet Union's effort to replace embargoed American technology. The Soviet Union simply could not trust the political climate to ensure a comprehensive working agreement. In effect, the Reagan administration had violated sanctity of contract in the cooperation. This explains the presence of the *Shelf 5* and the buying of ice-resistant technology from McDermott. Along these lines in 1984, the Soviet Union announced plans to build its own facilities for drilling rigs and the manufacture of offshore equipment.[130]

Present Developments

Since 1984, the Soviet Union has become more insular in its approach to the development of the region. Its primary concern in cooperating with Japan now is the liquid natural gas project at the Chaivo field. Even that seems to be on the wane. The issue here is similar to that which arose when a general agreement was negotiated over the 1970 to

1975 period. The Soviet Union is trying to get Japan to pay most of the $4.5 billion necessary for the field's development. The Japanese are reluctant to proceed too fast on this deal because many of its banks are unwilling to loan the money, owing to low fuel demand in Japan.[131] Another reason for the reluctance is probably the difficulties Japan has experienced with American equipment and license embargoes.

It would seem as though the Soviet Union has decided to develop the region more independently turning to Finland when necessary for advanced equipment. This is evidenced by the Soviet Union's asking Finland to participate generally in the development of the region, albeit with the Soviet Union controlling and deciding all stages of it.[132] The Soviet Union, for the meanwhile, is probably quite content to use the existing equipment it acquired from the Japanese and other Western powers. In the spring of 1986 the Soviet Union spud its first offshore well off the southern coast of Sakhalin near Aniva bay.[133]

Finally, in June of 1986, both the Soviet Union and Japan decided that, due to weak oil prices it would be forced to reassess their plans to develop the fields they discovered over the life of the project.[134] Essentially, the joint cooperation between the Soviet Union and Japan to develop the offshore waters of Sakhalin had come to an end. The liquid natural gas project at Chaivo is still underway but, it too, is experiencing problems with low energy prices and bank financing by the Japanese.

Lessons From the Sea of Okhotsk

One thing evidently clear from the Soviet Union's experience in the Far East is the nearly complete lack of Soviet technology capable of adequately developing the region. Further, dealing with Japan has shown that essentially the Soviet Union is purchasing and licensing American technology. American influence extended beyond rigs licensed to Japanese firms, as was shown by both the Carter and Reagan embargoes. Most of the West European firms the Soviet Union engaged to avoid American sanctions were forced to withhold equipment or expertise because of their numerous licenses from American sources.

Clearly, the Soviet Union had to reduce its dependence upon American technology, both directly and indirectly. In the development of the Arctic waters this would become a policy. Their need was to develop an ability to explore the offshore regions and engage as many non-American licensed offshore companies as possible. As a result, Finland, with its access to Norwegian equipment and its own experience in shipbuilding, would become a major player in the Soviet offshore industry. Countries developing North Sea oil were the primary targets for Soviet equipment deals, since these countries were developing their own technology independent of the Americans. Considerable caution would of course have to be exercised, particularly in the case of Norway, always one of the strongest supporters of the United States within the Coordinating Committee on Multilateral Export Control (COCOM), a part of the North Atlantic Treaty Organization (NATO).

The Soviet Union also seemed to be content to leave the work in the Sea of Okhotsk at the exploration stage. This further emphasizes the fact that exploration over production in the offshore waters would be a tolerable reality. Many critics viewed the deferment of the Sakhalin project as a setback in the Soviet petroleum industry. This can only be the case if one focuses on production operations. It is clear the Japanese were not eager to abandon the project. Indeed, they made a number of concessions throughout the negotiating and renegotiating process. It was the Soviet Union that forced a slowdown in operations. Because the majority of exploratory work had already been done to the Soviet Union's satisfaction, production was not a priority. Therefore, it would be premature to say the project represents failure; on the contrary, from the standpoint of exploration, it should be viewed as a success.

The Baltic Sea

Soviet offshore activity in the Baltic Sea has not been very substantial when compared to developments in the Caspian Sea and the Sea of Okhotsk. From the point of view of potential reserves in comparison to the Soviet Arctic, the region takes on even less importance. There has been only one find in the Soviet section of the Sea, although Poland

11. The Baltic Sea

discovered oil at two test well sites, and Sweden discovered petroleum onshore on Gotland island. West Germany has also discovered oil in its sector of the Baltic, but has not gone ahead with development plans. There are, however, at least three reasons for surveying development plans in the Baltic Sea.

First, while the area will surely never be a prolific oil producer, it does have the capacity to serve regional demands of the northwest European portion of the Soviet Union. This was one of the main reasons for the development of the Sakhalin fields offshore, as was pointed out previously. The drawback is the reserve potential of the region is nowhere near that of the Soviet Far East. As a result, the Baltic offshore development is seen more in the role of satisfying regional petroleum demands. It will often, therefore, take a lower priority to other offshore plans, such as those in the Caspian or Arctic waters. In part, this explains the relatively low activity in the region. The extent to which the Soviet Union dedicates resources to Baltic offshore development is an indication of the scarcity of exploration and production facilities for the offshore industry in general elsewhere. Probably the main use of the Baltic is to test rigs produced at the Vyborg yard.

Secondly, the Soviet Union has opted to cooperate with CMEA members rather than with Western partners as it did both in the Arctic and the Sea of Okhotsk. This cooperation is in the form of a consortium, Petrobaltic, established in 1976 by Poland, East Germany, and the Soviet Union. The problem here is little access to advanced technology as there was in cooperation with Japan in the Sea of Okhotsk or with Finland in the Arctic and Caspian seas. The use of Western technology from other regions to develop the Baltic Sea, is not an option since there is a scarcity of such technology. It is interesting to note the Soviet Union's lack of enthusiasm in Baltic offshore petroleum development, probably for this reason.

Third, the Soviet Union is concerned about the environmental impact of offshore petroleum in the Baltic Sea, which is relatively shallow and therefore very susceptible to permanent damage from oil spills. The recreational facilities in and around the Baltic republics are surely a factor in determining this.

Activities in the Baltic Sea

Geophysical work was done throughout the late 1970s and early 1980s. Little was done, however, in terms of exploratory drilling. One of the reasons for this slow pace of exploration, other than its low priority in the Soviet Union's offshore development plans, was due to complicated negotiations with Sweden over the island of Gotland. Sweden found oil on the island and have since been trying to determine the boundary between the Soviet Union and itself by demarcating the distance halfway between Gotland Island and the Soviet Union. The Soviet Union has not conceded this and instead attempted to determine the boundary as halfway between itself and Sweden.[135]

In 1976, East Germany began to draft plans for the development of wildcatting in the Baltic. While East Germany was assured of gas supplies from the Soviet Union up to the year 2000, it was concerned over its petroleum supplies. For sometime it has been the second-largest importer of Soviet petroleum, buying for example in 1975 an estimated 300,000 bd.[136] East Germany then was one of the primary forces behind the formation of Petrobaltic. Poland, as well, saw an opportunity to begin offshore development, largely in response to the harshness with which the energy crisis hit in the early 1970s.

Petrobaltic, probably on the heels of the Soviet purchase of equipment for its Astrakhan yard in the Caspian, began to order a number of foreign parts and rigs for offshore development. The first of these occurred in 1978 with a $49 million order to Rotterdam Shipyards (Netherlands) for a self-elevating, offshore oil-drilling rig.[137] This order was followed by one for a drilling platform to operate in 90 meters of water. The rig was constructed by Rhein-Schelde-Verolme (Netherlands) and was delivered in mid-1980. It is operating in the Polish sector of the Baltic at Przyladek Rozewie field just off Danzig.[138]

The consortium also used a Dutch-built jack-up to begin exploratory drilling in the Soviet sector of the Baltic as late as February 1984, quite some time after delivery of the jack-up to the Gdansk yard in Poland in 1980.[139] The lateness of the drilling in the Soviet sector suggests that the consortium, and most probably the whole Baltic development plan, was for the benefit of East Germany and Poland. East Germany was concerned over oil supplies. Poland experienced shortages of coal in the winter months of 1979 and 1980 and was actively seeking alternative energy sources.[140] From the Soviet Union's point of view, the Baltic Sea offshore exploration was bound to be of low priority.

Nonetheless, in November 1984, after 318 days of drilling to a depth of 2,800 meters, the first offshore well in the Soviet sector was found. It was named the 27-oi S"ezd KPSS.[141] The well is located off the port of Baltiisk in the Kaliningrad Oblast.[142] The well was drilled from a stationary platform and not from a jack-up as was thought at first. The test-pay was described as strong and the crude of high quality.[143] The find at Kaliningrad did give a small boost to the offshore industry. It allowed the Soviet Union to test one of its own semisubmersible rigs supposedly produced at the Uralmash yards (although it was likely outfitted there before final entry and built at Astrakhan).[144]

Exploration again resumed off the Polish coast in 1985 until the first well on the Polish shelf was discovered in 90 meters of water. This was followed by a second discovery around October 1985 of which little is known. There was a report in *Pravda* suggesting the Baltic Sea is a region with commercially productive petroleum.[145] It is difficult to say if this is the case, but if activity to date is any indication it does not seem likely.

In the last three years activity has been slow despite the finds in both the Polish and Soviet sectors. It was not until June 1986 that the Soviet Union began to build up its oilfield off of Kaliningrad. At this time a 1,210-ton jacket was towed from Klaipeda to the region, but production hole drilling has not yet begun.[146] There was an announcement in late 1986 the Soviet field would be coming on stream by 1989. The arrangements were that 50% of the production would go to the Soviet Union with Poland and East Germany each receiving 25% of the output. Probably each country had a similar arrangement with each taking half the output from its own wells.[147]

Most of the countries that border the Baltic have taken at least a cursory interest in the petroleum potential of the Baltic but none to such an extent to suggest the region will be of importance. The only sign of success was a West German find in its sector of the Baltic. In November 1984 Deutsche Texaco (FRG) and Wintershal (FRG) discovered petroleum off of Schwedeneck.[148] Little has been heard of this deposit since it was discovered, again suggesting a lack of interest in development. The West German discovery, however, does indicate there may be some kind of cooperation between either the Soviet Union and West Germany or between Petrobaltic (probably via East Germany) and West Germany in order to obtain West German technical expertise. To date, however, this has not occurred.

Lessons From the Baltic

While developments in the Baltic have been rare, one important consideration should be kept in mind. The Soviet Union's involvement with Poland and East Germany in Petrobaltic has largely been for the benefit of the latter two countries and not for itself. It is not really in the Soviet Union's interests to seek Polish or East German technical expertise since they have little to offer. The only use they may be to the Soviet Union in terms of technical cooperation is their access to Western technology in cases where the Soviet Union is unable to acquire some equipment. Probably East Germany is the main drawing card in this sense, given its special economic relationship with West Germany.

It must also be kept in mind that offshore oil exploration and development in the Baltic Sea was an early decision. Since it was undertaken, the emphasis has shifted to the Arctic and the deeper regions of the Caspian Sea, meaning that less investment funds are available for the area. Territorial disputes complicate work in the region, especially with Sweden, but also between East and West Germany in their regions of the Baltic. The Soviet energy program for the CMEA is currently characterized by constraint on the part of Soviet petroleum sales to that region. Heavy investment in offshore oil, where there is relatively little to be found, would contradict this current policy.

While the find off of Kaliningrad may have some regional importance, it can hardly be considered significant either in terms of potential reserves or as an indication of future developments. The slowness with which the Soviet Union has developed the region, therefore, suggests any Baltic Sea exploration and development project is really more in the interests of Poland and East Germany, with the Soviet Union providing the technology and expertise.

The Soviet Arctic

The sheer size of the Soviet Arctic shelf waters, comprising over 70% of the entire Soviet shelf, makes the region important. It is by far the most prominent area from the point of view of petroleum reserves. Much of the literature concerning the Soviet Arctic petroleum reserves discuss the potential of all six main seas which comprise the region. There are, however, only two regions of known hydrocarbon potential--the Kara Sea and the Barents Sea, which includes the Pechora Sea. The other seas--the Laptev, East Siberian, Chukotsk, and Bering--while speculatively hydrocarbon-rich, have had little or no Soviet activity. Until now there has only been American activity in the Bering Sea, which has has focused on undisputed waters for the most part in and around the St. George and Navarin basins.

An Overview of Soviet Arctic Offshore Petroleum Strategy

It is first necessary to explain Soviet strategy in the development of the Arctic. There are three factors that are of special note: first, the concentration on the development of the Barents and Pechora seas and the lack thereof on development of hydrocarbons in the Kara Sea; second, the emphasis on new forms of Arctic technology necessary to develop the region; finally, in response to these technological demands, the involvement of Western partners and the importing of Western equipment.

Why the Barents Sea?

The area that has experienced the most exploration activity since the Soviet Union seriously began to explore the hydrocarbon potential of the Arctic is the Barents Sea and its subsea the Pechora. This has been so for several reasons.

First, the Barents Sea presents less of a problem with

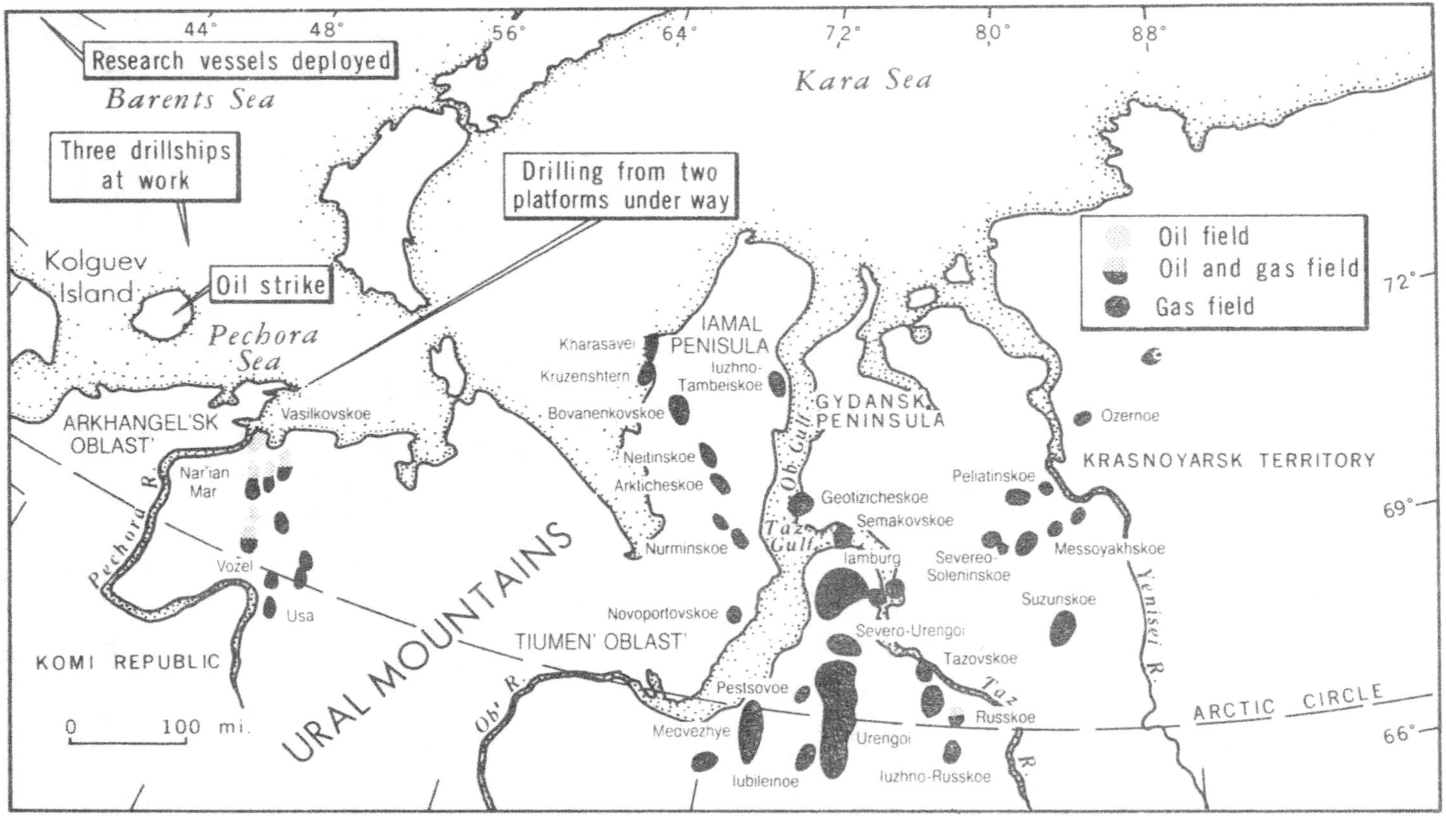

12. The Soviet Artic: Barents, Pechora and Kara Seas

ice cover. Most of the Soviet Arctic is covered with ice year round. The Barents Sea, like the Sea of Okhotsk, has ice cover only in winter; and, the area between Bjornoya, Norway and the Soviet Kola peninsula in the Barents is ice-free all year round. This has allowed the Soviet Union more freedom in exploration activity than has been afforded by the more hostile seas to the east. Further, this area near Norwegian waters enables the Soviet Union to rely in part on Norwegian seismographic surveys and hydrocarbon exploration of their sector of the sea. The extent to which the Norwegians make such information public could be problematical, but there have been agreements signed by both countries to share data on the region.

Second, the proximity to Norway of the ice-free region of the Barents has another advantage. Like Japan and the Sea of Okhotsk, the Soviet Union can rely on some forced cooperation from Norway because it is in both nations' interests to cooperate with one another. The difference here of course is that the Norwegians have developed their North Sea oil reserves to such an extent they can afford to take a slower approach to the Barents operation than Japan did to its offshore Sakhalin development. The Japanese had not seriously considered joint development of China's offshore waters when they entered into negotiations with the Soviet Union. This made Japan the junior partner in the Sakhalin development as it was in serious need of resources. In this respect, Norway is in a stronger bargaining position to hold out for a favorable development plan than Japan had been.

Third, the proximity of the potential Barents offshore fields are much closer to Soviet markets than the fields of West Siberia and those of the more recently discovered fields in East Siberia. Should offshore production become a reality, say, near Kolguev island and shipped onshore by tanker, the initial capital outlay for offshore technology may be less than developing and transporting petroleum from the remote West and East Siberian fields.

Finally, the large hydrocarbon potential of the Kara Sea is at present believed to be natural gas. Some observers argue that the Kara Sea may contain petroleum at lower depths, perhaps below 6,000 meters (see Chapter 3), but it would seem that Soviet planners have not recognized this. The main reason for this is that petroleum in the West Siberian basin has never been discovered north of 64 degrees latitude. Gas has been discovered in large quantities onshore, especially at the Urengoi and Iamburg fields. As was stated in Chapter 2, the problem of transporting gas via pipeline onshore is enough of a problem let alone offshore. There also is no need to spend resources developing offshore gas fields since Soviet gas production and reserves are more than adequate for the next three or four decades. The verdict on the Kara Sea, however, is by no means decided. James Clarke of the U.S. Geological Survey holds considerable optimism for the Kara Sea. He does, however, admit that to date extensive drilling on the Iamal and Giadi peninsulas have yielded only gas. No offshore wells have been drilled. His assessment that there are 6 bbl of oil reserves in the Kara Sea may very well be true; but considering the almost complete lack of Soviet activity in the region, it would seem the Kara Sea is not a priority offshore area to Soviet planners.[149]

Technological Constraints

Technological constraints are also a factor in the strategy of development. The issue is ice cover, low depths in places, and harsh Arctic winds. With the exception of the narrow strip between Norway and the Soviet Union in the north-west Barents, the region has a minimum of six to eight months of ice cover. The intention in the Sea of Okhotsk was to drill during the ice-free summer and moor ships and rigs in ice-free ports over the winter. In the Arctic the only ice-free port for the Soviet Union is the naval base on the Kola peninsula. This problem is compounded should there be foreign vessels with foreign engineers on board. The naval base is a high security zone.

The seabed of the Barents is folded in a complex way. Until recently, the Soviet Union did not have multi-folded seismographic equipment to carry out research on these types of geologic structures. Once the structures have been assessed, the problem is to find dynamically positioned drillships to drill exploratory wells. The installation of fixed platforms is out of the question due to freezing; and jack-ups, while in use, are not the most efficient way to drill exploratory wells if the intent is to limit the drilling season to a few months. The problem is also compounded by the cold in terms of worker productivity. *Ekonomicheskaia gazeta* has estimated that at -40 degrees Celsius, the productivity of Western Siberian drilling crews is reduced by 35% to 45%.[150] The same logic applies for drilling in the Arctic waters.

Drillship requirements are even more complicated. They must be able to withstand severe storms and wave action on the vessel so as to allow drilling to take place. An essential component of these ships is, therefore, a highly complex computer-operated propeller system, something that has only been developed since the early 1980s in the West.

There is a further technological problem presented by safety and support systems. The Soviets are quite ill-prepared to deal with blowouts at wells offshore. One such example was in the Caspian Sea, where a blowout at an offshore gas well lasted for eight months, largely because there were no blowout preventers on the rig.[151] One can imagine if this was a difficulty in the temperate zones of the Caspian, in Arctic conditions it would be horrendous.

The issue of support for rigging is also a concern. The Soviet Union has found it necessary to buy a number of special ice-resistant sleeves from the French for the Soviet rigs as well as de-icing equipment for the apparatus on board the rigs. Underwater maintenance in the Arctic is also an issue the Soviet Union has only recently faced in its purchasing of diving bells for Arctic conditions. Once a rig is operating, there is the additional problem of employing icebreakers for tankers shipping petroleum onshore. If the issue is laying pipe, then icebreakers must be employed also. Until recently Soviet icebreakers were inadequately equipped to deal with such concerted efforts.

These can hardly be viewed as criticisms of Soviet preparation for Arctic development. The only other country that has undertaken large-scale Arctic plans of this magnitude is Canada. Its results have been limited to date, and its problems numerous and similar. Nevertheless, despite the Soviet Union's apparent lack of technology for offshore development, by 1986 the Soviet Union had become the leader in Arctic hydrocarbon exploration. In many ways this was due to a concerted effort to systematically develop the offshore reserves. Inclusive of this is an assessment of the needs of the region, then making the appropriate purchases from the West.

The Involvement of Western Firms

In the Soviet Union's efforts to develop the Arctic, the Soviet Union has attempted to engage as many firms as possible without allowing too many foreigners into the region. The Soviet Union is surely more cautious than when it dealt with Japan. The Soviet Union understands the extent to which American technology dominates the international offshore petroleum industry. It is certainly not beyond the Soviet Union's abilities, however, to find the necessary technology elsewhere; it is just not as easy. For example, in a report to the Subcommittee on International Economic Policy and Trade of the Committee on Foreign Affairs of the United States House of Representatives, Armco (U.S.) drew up a list of parts that went into a deep-water semisubmersible rig which had been sold to the Soviet Union. The finding was that of the 25,000 essential parts and the million or more individual pieces that comprised the rig, all were available outside the United States.[152]

The Soviet Union's actions to date suggest therefore that it is going to engage as many non-American partners as possible. Finland, not surprisingly, is the main trading partner in the Soviet Union's Arctic strategy. Finland has always played a special role in Soviet industrial development. In 1985, 75% of all crude imports to Finland were from the Soviet Union, amounting to nearly 1.7 billion rubles.[153] Similarly, in 1985 the Soviet Union's largest share of its imports from Finland were in machinery and equipment, estimated at over 1.4 billion rubles.[154] As has been seen in the discussion on the Caspian and Okhotsk seas, Finland has been an integral part of the Soviet Union's offshore development. In the case of offshore Arctic development, this role is especially pronounced. The proximity of Finland, like Norway, is also a reason for this. Finland's special status with Norway and its free access to West European markets are added attractions for the Soviet Union.

Early Plans for Soviet Arctic Exploration

Plans to explore the Soviet Arctic began around 1976 with the discovery of hydrocarbons on the Kola peninsula near the Barents Sea. The Kola peninsula hole in 1976 was over 6,000 meters deep when hydrocarbons and helium were discovered. The hole, located 80 kilometers north of Murmansk, is near the Norwegian border.[155] The discovery forced the Norwegians to begin dealing with the Soviet Union on the issue of the boundary between the Soviet Union and Norway in the Barents Sea.[156]

The line runs northward from the base of the Kola peninsula through the straits between Franz Joseph island and Spitsbergen. The territory of Spitsbergen was declared Norwegian in 1946, but under a United Nations charter some 60 nations were granted mining rights to the island. The only two countries to exercise these rights are Norway and the Soviet Union. The discovery of hydrocarbons on the Kola peninsula, therefore, probably jolted both nations into cooperation. At present, only coal is produced at Spitsbergen (and at that, at a loss for Norway and probably for the Soviets, although they claim otherwise).[157] If petroleum was discovered on the shelf of the island, however, complications over property rights would surely emerge.

One finds in the discussion of the development of the Arctic exploration activities of the Soviet Union the boundary dispute between Norway and the Soviet Union is mentioned only rarely. This is probably due to the fact there is no need yet to squabble over hydrocarbon reserves that will not be fully developed for more than a decade. It is a situation like the one the Soviets experienced with Japan in the Far East. It is better to cooperate when the stakes are not high, this with hope of evolving an agreement over mineral rights. This was probably both the Soviet Union's and Norway's intent.

In 1976, the Soviet Union began to open inquiries with several countries about the acquisition of mobile offshore rigs. France, Finland, Norway, and the United States were all consulted. The first serious negotiations to directly involved the Arctic, however, were with Norway and Finland. The Soviet Union was interested in acquiring two of the Aker Group (Norway) H-3 semisubmersible rigs.[158]

No information on this deal has been heard of since. The Soviet Union eventually bought semisubmersibles from Finland for use in the Caspian Sea. But the dealings with Norway had been initiated, and the fact that there was now a potential border dispute over petroleum in two regions (north from the Kola and on the Spitsbergen shelf), meant that Norway and the Soviet Union would be partners no matter the situation. This was to be of some importance later on in 1982 when the two sides once again sat down to decide a plan for Barents Sea development.

In the meantime between 1976 and 1979 the Soviet Union seems to have taken its time to develop a strategy of exploration activity. It may have been the Soviet Union was simply too preoccupied with developments in the Caspian Sea and the Sea of Okhotsk to begin contemplating the numerous problems involved in Arctic offshore oil exploration. There was probably some degree of waiting for the results of exploration activity in the more temperate waters of the Caspian Sea in order to obtain a gauge for what constitutes a successful offshore development, and in the Sea of Okhotsk, to find out how much equipment the Soviets could acquire via their special agreement with Japan. Events which occurred after 1979, however, suggest the Soviet Union was in fact planning in these years. A number of references are made to the Soviet Union's geophysical work in the Pechora Sea and in the southern Barents Sea.[159]

The Soviet Union undertook some important deals for infrastructural equipment. It licensed from Pusnes Mekaniske Verksted A/S (Norway) in 1978, equipment to assist in the production of its cold-weather mooring systems. These were to be used on the anchoring equipment for semisubmersible rigs. At the same time a deal was reached with Elf Aquitaine (France). The French firm would provide financing and technology for the Soviet Union's Barents Sea exploration.[160]

A number of onshore tests were initiated at this time that tended to raise more hope about potential hydrocarbon discoveries offshore. There have been four onshore tests of note: one on Kolguev island in the Pechora Sea, one on the Iamal peninsula, and two on Franz Joseph island. The test on the Iamal mainly revealed gas-pays, but the other three wells revealed evidence of petroleum.[161]

At this time there was also consideration to limit the costs of offshore petroleum production. This was to be done by cluster drilling from deviated holes from the shore. In essence it was a more advanced type of slant drilling of the type done from Sakhalin island in the 1920s. Deviated holes were also to be drilled from dredged islands and from concrete platforms installed in shallow waters.[162] The prospects of these developments, however, are not mentioned after 1979, probably because they were designed for the Kara Sea, which is quite shallow. When geophysical work revealed the predominance of gas here and strong evidence of petroleum in the deeper Barents Sea, these plans were likely abandoned.

Exploration Increases

A benchmark era in the Soviet Union's exploration plans of the Arctic is 1982 to 1983. It was in 1982 that the first of three drillships began to arrive from Finland. Major construction plans were undertaken at Vyborg to convert the yard to build semisubmersible rigs for Arctic offshore development, an improvement over the abilities of the Baku, Astrakhan, or Narimanov yards in the Caspian region. Complementing this was the more complex geophysical work undertaken by the Institute of Oceanology throughout 1982 and 1983. Finally, it was in late 1983 the Soviet Union began to develop drillship technology on its own, probably by a process of reverse engineering. This is something not often mentioned in press reports on the Soviet Arctic developments, but must surely be viewed as a most important factor in the Soviet Union's Arctic offshore development.

Soviet exploration plans in the Arctic can therefore be broken down into three phases that overlap in time. These shall be dealt with in turn for the remainder of this chapter.

First, the delivery of the drillships from Finland was extremely crucial to exploration plans. Following these deliveries, the development and design of a Soviet drillship heralded a new era in the development of the region. The discoveries from these ships were not numerous but were significant. The work of these drillships was then complemented by advanced geophysical research vessels, completely of Soviet design and origin, with the exception of the *Chaivo*, brought in from the Sakhalin development.

Second, as a complement to the domestic research ship developments, the Vyborg yard was designed for semisubmersible construction and the deployment of the next four

Shelf rigs. This phase began at the same time as the delivery of the drillships and was completed at about the time the drillships made their discoveries.

Finally, the cooperation agreement signed between Norway and the Soviet Union in 1983 to develop the Barents Sea initiated a whole series of negotiations between the Soviet Union and other Western partners, ranging from France and Italy to Brazil and Canada.

The Delivery of the Drillships

The Soviet Union placed an order with Rauma-Repola Oy (Finland) for three dynamically positioned drillships to be delivered between 1981 and 1982. At the time of signing, the order was valued at $225 million. In the spring of 1982, the first of these drillships, the *Valentin Shashin* was delivered. It was based on a design by IHC-Gusto Pelican (Netherlands). Rauma Repola Oy subcontracted much of the work out to RSV Gusto Engineering (Netherlands) and then undertook redesign work and modifications at its own shipyards.[163]

The Soviet Union displayed some considerable foresight in this deal. It did not place an order for a more advanced ice-resistant jack-up or a semisubmersible rig with the Finnish firm; instead, it placed an order for one of the most advanced, dynamically positioned drillships in the world. The *Valentin Shashin* was towed to a location approximately 70 degrees north and 42 degrees east, which would place it on a hydrocarbon-rich structure known as the Murmansk high (see chapter 3).[164] It was here that the first well was spudded from the *Valentin Shashin*, the test-pay of which is unknown.[165]

The *Valentin Shashin* was then followed by two other drillships--the *Viktor Muravienko* and the *Mikhail Mirchink*. These two drillships are similar to the *Valentin Shashin*. All three are designed to drill in waters of up to 300 meters and can drill to depths of 6,000 meters.[166] Little is known about the *Viktor Muravienko*, except that it spudded a well near the other well found by the *Valentin Shashin*.[167] It is supposedly either drilling in the ice-free portion of the Barents Sea near Norwegian waters or in the Kara Sea.

The *Mikhail Mirchink* joined the other two ships late in 1982, but was reported to have left the area to join the Far Eastern Marine Oil and Gas Exploration Expedition based at Korsakov on Sakhalin island.[168] There were plans to have the vessel conduct research and surveys off of the Kamchatka peninsula in the sensitive Western waters near Alaska, these reports, however, have never been confirmed.[169] At last reports, the *Mikhail Mirchink* was damaged while operating off the coast of Vietnam and was being repaired by Keppel Shipyard Co. (Singapore) at a cost of $2.8 million dollars.[170]

Problems began around the summer of 1982 for the *Valentin Shashin*. It was reported the ship had been severely damaged due to the inability of its all-Soviet crew to operate the vessel. The vessel had its risers sheared off and its frame badly contorted. This was unusual since it was operating in only 82 meters of water. The ship was then reportedly towed to the Rauma-Repola Oy yards for repair. In March 1983, the ship was reported as leaving the Rauma-Repola Oy yards and was once again placed in the western Barents Sea near Norwegian territory. It was then reported to have left that location, and nothing has been heard of it since.[171]

One of the most interesting features of drillship activity in the Arctic is the deployment of the first Soviet-built drillship: the *Diorit*. The ship was constructed at the Iaroslavl' Shipyards in western Soviet Union. It is known that the *Diorit* is able to drill in only 100 meters of waters, nearly 200 meters less than its foreign counterparts.[172] This drillship, due to its limited depth capabilities, is probably going to drill on the shelf of Kolguev island where petroleum was discovered onshore in early 1983.[173]

Geophysical Research Vessels Operating in the Arctic

Complementing the work of the drillships at this time was the presence of several geophysical research vessels. Most of the geophysical work carried on in the southern portion of the Pechora and Barents seas was undertaken by two research vessels built under the direction of the Soviet Institute of Oceanology. These were the *Professor Shtokman* and the *Akademik Kurchatov*.[174]

In 1983, the *Valentin Shashin* was joined by another research vessel in the western Barents known as the *Spurt*. This Soviet-built vessel operated by the Arktikmorneftegaz-razvedka was being used to train divers in the use of manned submersible crafts at depths of 300 meters. It was reported in the beginning of October 1983, a crew descended from the *Valentin Shashin* to the wellhead drilled earlier that year to 250 meters, constituting a record for the Soviet Union.[175] The diving bell is able to stay down for as long as 30 days and is used in repairs of underwater bore holes.[176] These manned bells, known as the *SM-170* and the *Moana*, were supplied by Comexindustrie (France) and are reportedly being operated with the assistance of French engineers.[177] If this is the case, they would be the only Western operatives in the Soviet region working on Soviet offshore Arctic exploration. In the summer of 1984, the *Chaivo* from the Sea of Okhotsk, a Finnish-built research vessel known as the *Akademik Golitsyn* and the *Professor Shtokman* were all sent north to operate in the Barents Sea near Norwegian waters.[178]

There are recent reports the Soviet Union has received a Soviet-designed research vessel, the *Dimitrii Nalivkin*, from Finland in March of 1985.[179] This is believed to be the first of two on order from Rauma-Repola Oy, worth in excess of $23 million.[180] While the information on the drillships and research vessels is scanty at best, there are three observations of note to be made.

First, the Soviet Union appears to have gone about exploration in the right manner, gathering as much information as possible before employing the extremely expensive jack-ups and semisubmersibles to expand well development. This is implied by having research vessels operate in the Barents at the same time as the drillships. The Soviet effort seems more organized and thorough than many Western critics admit.

Second, however, is the unfortunate rationing of exploration technology that has to occur in order to conduct such an exploration program. This is evidenced by the numerous locations of the *Mikhail Mirchink* and the fact that the *Valentin Shashin* was not replaced when it encountered difficulties in 1983. Further, the fact that it had to be towed to the Rauma-Repola Oy yards for refitting and that the *Mikhail Mirchink* had to be repaired by a Singapore firm says that the Soviet Union, despite smart foreign buying, has failed to keep pace with the technology it has imported from the point of view of maintenance.

Finally, it is interesting that the Soviet Union is now slowly trying to replace the foreign technology. The presence of the Soviet research ships and the *Diorit* drillship is evidence of this. Yet, one can assume their technical complexity is not as advanced as the Finnish-built drillships, otherwise maintenance on the *Mikhail Mirchink* would have been undertaken by Soviet engineers. Further, the most recent research vessel, the *Dimitrii Nalivkin*, though of Soviet design, had to be constructed in Finland.

The Vyborg Semisubmersible Yard and the Expansion of Finnish Cooperation

In 1982, the Soviet Union began to develop its first deep-water semisubmersibles for Arctic conditions. Development plans were announced in 1982.[181] They entailed upgrading the Vyborg shipyard in the Soviet Union near the Finnish border, to construct both semisubmersibles and, eventually, ice-resistant jack-ups. This was brought about for the same reason as the development and equipping of the three semisubmersible yards in the Caspian Sea. The Soviet Union realized it would require substantial domestic experience in technological developments for offshore drilling if its industry had any hope of long-term survival.

The original plan was to develop four Shelf class semisubmersibles of a more complex and weather-resistant design than the three semisubmersibles operating in the Caspian Sea. While it is claimed these are exclusively Soviet in construction, there is evidence of some important foreign contributions. The three Shelf rigs operating in the Caspian Sea are probably built from the hulls of the Finnish Pacesetter units the Soviet Union purchased from Finland in 1983. The ones built at the Vyborg yard for Arctic use also have some foreign input and may be based on the Pacesetter design. There is evidence Finland had some part to play in these Arctic semisubmersibles in a report which refers to a semisubmersible, the *Vyborg I*, which "...is to undergo final outfitting before being turned over to the Soviet government in 1983,"[182] at the same time the yard was under construction, suggesting that the Soviet Union copied one of Finland's advanced Arctic semisubmersibles.[183] It is assumed all of the Shelf rigs are generally of the same abilities rated to over 6,000 meters of drilling in about 200 meters of water, all presumably ice-resistant.

In May of 1985, the *Shelf 4* was tested in the Baltic Sea, then released to Murmansk to operate in the western portion of the Barents Sea.[184] There have been reports that the *Shelf 4* was in fact drilling on the shelf of Kolguev island.[185] This is quite probable given the onshore find at Kolguev. At last reports the *Shelf 5* was still at the Vyborg yard, but was scheduled to drill in the Sea of Okhotsk and was to join the *Okha* rig north of Odoptu. It is also believed the *Shelf 5* may be operating with the *Mikhail Mirchink* off the coast of Vietnam.[186] These reports, however, have not been confirmed. There are two more semisubmersibles: the *Shelf 6*, which was to have been delivered to the Barents Sea sometime in 1986; and the *Shelf 7*, scheduled for delivery sometime in 1987.[187] More recent reports have, however, confirmed that the *Shelf 6* left the Barents Sea and has been drilling since May 1988 at a new South China Sea location, Danhung, off the Vietnamese coast.[188] The *Shelf 7* is reportedly drilling in the Barents Sea but these reports are unconfirmed.

The extent of foreign equipment on the Shelf series is not known. The design was probably Finnish, but the equipment was more than likely an assortment of gear from the Soviet Union's purchases abroad and developments at the Astrakhan and Baku yards and perhaps at the Uralmash yards. It is known, however, that Seaforth Maritime (Scotland) supplied advanced saturation technology worth around $10 million for the *Shelf 4*.[189] The foreign component, while unknown, is typically important. This is certainly the case when the Soviet Union placed an order for two more jack-ups from Finland in 1983.

The two jack-ups, the *Kolskaia* and the *Sakhalinskaia*, are two of the largest jack-ups in the world.[190] Valued at $180 million, the jack-ups are rated to drilling depths of over 6,500 meters and can drill in 100 meters of water. Each has three legs about 140 meters long and each weighs well in excess of 20,000 tons.[191] The jack-ups were to be delivered in July and November 1985.[192] While the firm to build and deliver the jack-ups, Rauma-Repola Oy, was seen as efficient and dependable, the Soviet Union was concerned the demands of the Arctic on foreign technology still meant the Soviet Union had to diversify their import sources.[193]

Over this period of time a series of contracts between the Soviet Union and several Finnish firms and other third parties via Finland were signed. Saga Petroleum (Finland) is reported to have offered the Soviet Union some kind of Arctic offshore technology, but nothing has been confirmed.[194] Valmat (Finland) undertook a cooperation deal with Foster Wheeler Petroleum Development (U.K.) to acquire advanced Arctic offshore drilling technology. With its bidding on various Arctic exploration projects, Valmat has already expressed an interest in selling this technology to the Soviet Union. The form this will take is through the sale of production platforms, topsides, and modules.[195] Kone (Finland) provided a catamaran vessel to Sudoimport, presumably for the development of the Vyborg yard. The *Titan-1* is the largest in the world, has a hoisting capacity of 600 tons and is a center-pillar type. At last reports Kone was to build three more for Sudoimport.[196]

This in no way meant the Soviet Union was trying to move away from Rauma-Repola Oy. In 1986, the Soviet Union ordered from the company, leg covers and elevating systems for the two jack-ups Rauma-Repola Oy was constructing for the Soviet Union, each valued at $45.9 million. Further, the Finnish firm negotiated the delivery of a whole rig construction site to enhance the Vyborg yard. Like Saga Petroleum, Rauma-Repola Oy decided to enter into a joint venture to gain access to higher technology. In mid-to-late 1985, Rauma Repola Oy signed an agreement with Brown and Root Co. (U.K.) to produce a floating production unit and a mobile Arctic drilling/production rig.[197] This would most likely be part of an improved package of offshore technology Rauma-Repola Oy would eventually sell to the Soviet Union. The other Finnish firm to become involved in the jack-up construction was Parkano Engineering Works, which is to be responsible for the construction of the 37 elevating units on the two jack-ups.[198]

Cooperation with Norway and Increased Foreign Participation

The expansion of foreign buying was complemented by a widening of relations in the form of joint development plans with Norway. Two events were of key importance in initiating this cooperation.

The first was the damage of the *Valentin Shashin* drillship in the summer of 1982. If the damage resulted from mishandling the drillship rather than adverse Arctic conditions, then the Soviet Union probably saw a need to enlist foreign help for training and ship-exercise maneuvers. Cooperation with Norway would afford the Soviet Union access to the one country with the most experience in Arctic offshore exploration.[199]

Second, and probably more important was the presence of three Soviet research vessels and two drillships in the western Barents near disputed waters. It was to Norway's advantage to take a part in the development of these reserves in order to claim both countries were jointly responsible for exploration work and, therefore, to be in a better bargaining position should the issue become a boundary dispute. The logic behind this type of cooperation is similar to the logic behind the earlier Japanese-Soviet cooperation in the Sea of Okhotsk.

In April of 1983, the Soviet Union requested from Norwegian Petroleum Consultants a plan for the joint development of the west Barents Sea. This followed after lengthy negotiations between the Norwegian company and Sudoimport.[200] Eventually, Norwegian Petroleum Consultants joined six other companies to form BOCONOR a Norwegian consortium similar to SODECO. This document, produced by Norwegian Petroelum Consultants and known as the Master Plan outlined cost estimates as well as goods and services that could be provided by Norway in the development of the Barents Sea.[201] The plan also detailed the use of fixed platforms (presumably ice-resistant) and sub-sea production units,[202] which would probably assist the divers operating the Comexindustrie (France) manned submersible crafts operating around the *Valentin Shashin.*

There was some controversy over whether the consortium arrangement would be threatened by the boundary dispute in the Barents Sea between Norway and the Soviet Union.[203]

It would seem more likely that just the opposite is true. The agreement in part exists because of this boundary dispute, not in spite of it. The Soviet Union has shown signs recently of benefiting from the venture. In 1985, the Soviet Union considered granting licenses to begin exploring the region to several Norwegian firms.[204]

The cooperation between Norway initiated a number of other foreign contacts that the Soviet Union sought out individually. One of the first of these was with Italy. Three Italian firms became involved in negotiations with the Soviet Union--Sirtec, ENI, and IRI.[205] Italian companies had a part to play in the development of the Urengoi-Pomary-Uzhgorod gas line when nearly $4.5 million worth of large diameter pipe and compressor equipment were provided. It is believed Italy will provide expertise and equipment for offshore platforms and underwater pipelines at very low depths.[206] This piping may have already been put to use. There were reports the Soviet Union was already shipping token volumes of petroleum to Arkhangel'sk from the eastern sector of the Barents Sea.[207]

A more recent engagement with a foreign country is the Soviet Union's countertrade deal with Brazil. Valued at $750 million for a four-year term, Nobara Sociedade de Mineracao Comercio & Industria (Brazil) is to provide the Soviet Union with 10 offshore oil-drilling platforms (presumably of Brazilian make) in return for Soviet petroleum. The Brazilian state oil company Petrobas is to take shipment of the petroleum.[208] As there are plans for at least five more rigs in the Arctic and up to 21 more in the Caspian by 1990, it can be assumed some of these rigs are bound for the Arctic.[209] Brazil's countertrade deal has not yet been confirmed. Given its size, it may be the deal never went through. Nonetheless, the fact the Soviet Union would consider such a large countertrade arrangement with a less-than-usual trading partner is evidence the Soviet Union is more than interested in developing its Arctic offshore waters.

The only northern neighbor not directly involved to a large extent with the Soviet Union's Arctic plans, other than the United States, is Canada. This is probably because of Canada's compliance to COCOM regulations and the generally similar line Canada takes with the United States vis-a-vis the Soviet Union. There have been negotiations between Canada and the Soviet Union on a number of Arctic issues, however, for several years. To date nothing has been resolved with regard to mineral or boundary disputes. The Soviet Union has stressed, however, if negotiations are to progress between the two countries, Canada will have to buy more goods from the Soviet Union. In 1985 for example, Canada imported only 18 million rubles worth of Soviet goods, but exported almost 950 million rubles worth of goods to the Soviet Union. In a speech in Calgary in June of 1985, Valerii Savinov, (a Soviet trade representative), stressed if Canada does not do something about this trade imbalance Canada may find itself excluded from the Soviet Union's Arctic exploration and development plans.[210]

The only deal that seems to have been struck between Canada and the Soviet Union in Arctic development is between Versatile Corp. (Canada) and GKNT (Soviet State Science and Technology Committee). Versatile will provide expertise to the Soviet Union in the construction of icebreakers, experience Versatile gained while building breakers for the Beaufort Sea. In return Versatile will receive technical information from Soviet Arctic survey vessels. It is hoped the cooperation will lead to sales of icebreakers to the Soviet Union.[211] The chances of this occurring, however, are probably slim. The Soviet Union has fairly advanced icebreaker technology, and what they do not have they can get from Finland. It would be more probable that Canada will participate in the Soviet Union's offshore development via joint ventures with Finland and Norway.

This raises an interesting point in terms of territoriality. By engaging Japan and Norway in the development of offshore reserves in disputed waters the Soviet Union initiated, on the whole, healthy discussions on where boundaries would be drawn. In Canada's case, the issue is one of sovereignty in the Arctic region. For this reason Canada can be expected to try and establish a presence in the Arctic and therefore, like Norway and Japan, may find itself forced to cooperate with the Soviet Union in petroleum developments.

It would seem that those countries which have maintained a reasonable two-way trade with the Soviet Union will benefit most from Arctic development. Finland has been the mainstay in East-West trade in advanced offshore equipment for the Soviet Union, but even Brazil imported over 70 million rubles of goods from the Soviet Union in 1985, while Norway imported 104 million rubles worth of goods in 1985 and France well over a billion rubles in the same year.[212]

The Soviet Union has, despite claims to the contrary that its Arctic developments were significantly slowing down, entered into another large-scale deal with Finland for more Arctic offshore equipment. In January 1987 it was reported that Neste (Finland) agreed to import 11.5 million metric tons of petroleum from the Soviet Union in a bilateral trade protocol. Neste is to reexport two million tons of petroleum and import 58.3 bcm of gas as well. In return the Soviet Union will receive a geotechnical research ship (2,000 dead-weight tons), an offshore drilling and transport ship, two accommodation ships, an underwater exploration and rescue service vessel, two 5,000 dead-weight ton, Arctic-class product tankers, jack-up platform legs and a jack-up unit.[213]

This may be the beginning of a new wave of foreign buying in the region. It was reported in August 1986 that the Soviet Union's Arctic Offshore Oil and Gas Exploration Trust had carried out only 46% of its assigned work in the last FYP.[214] Combine this information with the Soviet Union's announced plans in the twelfth FYP to double offshore exploratory drilling, and much more foreign buying will probably result.[215]

Notes

1. A.A. Meyerhoff (1980), p. 168.
2. Marshall I. Goldman (1980), p. 126.
3. I. Khvedchuk, "Put' k morskim mestoroshdeniiam," Pravda, (April 3, 1978), p. 3.
4. A.B. Suliemanov, R.P. Kuliev, E.I. Sarkisov, K.A. Karapetov, Ekspluatatsiia morskikh neftegazovykh mestoroshdenii, Nedra (Moscow, 1986), p. 7.
5. I. Khvedchuk (1978), p. 3.
6. I. Khvedchuk (1978), p. 3.

7. I. Khvedchuk (1978), p. 3.

8. V. Shashin, Neftianoe khosiaistvo, no. 3 (1974) p. 4. Cited in T. Gustafson, (mimeo, 1986) p. 34.

9. T. Mukhamed-Rakhimov "Put' k kaspiiskoi nefti", Izvestiia, Sept. 7, 1983, cited in T. Gustafson (1986), p. 19.

10. T. Gustafson (1986), p. 19 ff. 8.

11. An important Soviet source has recently come to my attention in which it is corroborated that indeed the Ministry of Gas is fully responsible for exploration. It goes on further to outline the complicated structure of its organisational tasks and some of the investment requirements it has undertaken to date. See A.I. Perchik, Ekonomika osvoeniia morskikh mestoroshdenii nefti i gasa, (Moscow: Nedra, 1987), especially pp. 33-52 and 84-86.

12. H.O Bergsen A. Moe, W. Ostreng, The Fridtjof Nansen Institute, Oslo, "In Search of Oil and Security: Soviet Interest in the Barents Sea", Noroil (August 1986), p. 109.

13. A.B. Suliemanov, R.P Kuliev, Eh.I. Sarkisov, K.A. Karapetov (1986), p. 13.

14. A.B. Suliemanov, R.P. Kuliev, Eh.I. Sarkisov, K.A. Karapetov (1986), p. 13.

15. Joseph P. Riva (1976), p. 482.

16. A.B. Suliemanov, R.P. Kuliev, E.I. Sarkisov, K.A. Karapetov (1986), p. 14.

17. "Soviets' Caspian Oil Flow to Rise Slowly", Oil and Gas Journal (March 10, 1986), p. 32.

18. A.B. Suliemanov, R.P. Kuliev, E.I. Sarkisov, K.A. Karapetov (1986), pp. 12-13. See also N.A. Ermenko et. al. in Joseph P. Riva (1976), p. 489.

19. A.B. Suliemanov, R.P. Kuliev, Eh.I. Sarkisov, K.A. Karapetov (1986), p. 13.

20. Joseph P. Riva (1976), p. 489.

21. Soviet Geography, vol. 25 (Dec. 1984), p. 777.

22. Oil and Gas Journal (March 10, 1986), p. 32.

23. Soviet Geography, vol. 25 (Dec. 1984), p. 777.

24. Oil and Gas Journal (March 21, 1983), p. 70.

25. "Soviets Press 28th April Development", Oil and Gas Journal (Dec. 10, 1984), p. 53.

26. Baku VYSHKA (May 12, 1985), p. 1, translated in JPRS, UEN-85-017.

27. Baku VYSHKA (June, 1985), p. 1 translated in JPRS, No. UEN-85-017, and Offshore (May, 1986), p. 110.

28. Oil and Gas Journal (May 4, 1987), p. 63.
29. "Russian Strikes Dot East Caspian Sea," Oil and Gas Journal (Aug. 2, 1976), p. 74.
30. Oil and Gas Journal (Aug. 2, 1976), p. 75. Little evidence that any oil was found there is corroborated by the fact that there have been no reports of production from Zapadno-Okarern. In A.B. Suliemanov, R.P. Kuliev, E.I. Sarkisov, K.A. Karapetov (1986), no mention of this field is made in discussing the past and future development of the western Turkmenistan shelf, only LAM bank, Zhdanov bank and Livanov banks (presumably east and central), are mentioned, see pp. 18-21.
31. This information comes from Isabel Gorst, Petroleum Economist (April, 1984), p. 145. It is doubtful however if she is correct in assuming that Rakushechnoe-more will be productive, since A.B. Suliemanov, et. al. (1986) do not mention the field for future development.
32. Baku VYSHKA (Jan. 26, 1985), p. 1, translated in JPRS, UEN-85-011.
33. Baku VYSHKA (May 30, 1985), p. 1, translated in JPRS, UEN-85-017.
34. Baku VYSHKA (April 5, 1985), p. 1, translated in JPRS, UEN-85-016.
35. R.I. Asdov, Gazovaia promyshlenost', no. 6 (June, 1984), pp. 44-45.
36. Baku VYSHKA, no. 6 (June, 1985), p. 1, translated in JPRS, UEN-85-017.
37. A.B. Suliemanov, R.P. Kuliev, E.I. Sarkisov, K.A. Karapetov (1986), p. 18.
38. A.B. Suliemanov, R.P. Kuliev, E.I. Sarkisov, K.A. Karapetov (1986), p. 18.
39. Some sources claim that the rigs were being designed to operate in 70 meters of water, see Soviet Geography, vol. 25 (Dec. 1984) p. 779. Suliemanov, et. al. (1986) and The Soviet Union: Energy Review, The Economic Intelligence Unit, no. 4 (1983), maintain that the intent was to design rigs that could operate up to 60 meters.
40. Noroil (Aug., 1986), p. 110. This report is a much later one claiming these rigs are operating. In an earlier EIU report, "Soviet Energy Review" No.4 1980 p. 96, David Wilson claimed these three rigs were classified as obsolete. It is quite probable that the rigs were retired when the "Baku" was deployed, but were redeployed to meet increased

drilling demand, after the EIU report was released.

41. Office of Technology Assessment (OTA), Technology and Soviet Energy Availability, (Boulder, Colorado: Westview Press, 1982) p. 216. A considerable amount of controversy surrounds the "Baku". According to Joseph P. Riva (1976), p. 496, and Ocean Weekly Report (July 5, 1976), the "Baku" collapsed and sank shortly after drilling its second hole in 1976. David Wilson in an EIU report maintains that the "Baku" in fact drilled over 3,000 meters by the fall of 1978 at the Iushnaia-2 deposit. H.O. Bergsen, A. Moe, and W. Ostreng, Noroil (Aug., 1986), p. 110, corroborate Wilson's report by stating that as of summer 1986 the Baku was still operating in the Caspian Sea. The rig which collapsed was more than likely the "60 let Azerbaidzhana" as was admitted by the Soviet press in April 1987. If, however, the "Baku" collapsed in the year Riva suggests, 1974, then another rig must have also collapsed, since the "60 let Azerbaidzhana" was not constructed until 1978.

42. Joseph, P. Riva (1976), p. 480.

43. D. Wilson, "Soviet Energy", EIU, no. 4 (1980), p. 96.

44. Noroil (Aug. 1986), p. 110.

45. Sotsialisticheskaia industriia (May 17, 1985), p. 3, translated in JPRS, UEN-86-018 p. 66, claimed that the "60 let Azerbaidzhana" was drilling at Altai More as late as 1987, recent admissions however disprove this. This answers a question raised by Joseph Riva (1976) who claimed that the "Baku" had actually sunk shortly after operations began, this is untrue, but Riva's confusion is understandable given that the Soviet press did not admit the accident of the "60 let Azerbaidzhana" until at least 10 years after the fact. See Oil and Gas Journal (April 6, 1987), p. 73.

46. See EIU, no. 4, 1980 p. 96 and Noroil (Aug., 1986), p. 110.

47. See Joseph P. Riva (1976), p. 496, and Offshore (Sept. 1982), p. 19. The spelling of the pipe-laying barge "Suleiman Yeyirov" is taken from Oceanic Weekly Report (Aug. 1974), vol. 8, no. 47, p. 3.

48. Soviet Business and Trade, vol. 8, is. 5 (July 18, 1979), p. 2. The evidence that this activity was in the Caspian Sea was derived from a reference by Marshall I. Goldman, The Enigma of Soviet Petroleum: Half Empty or Half Full, (London: George Allen and Unwin, 1980), p. 127, to Geo-

source. Goldman does not detail the operation however, he merely mentions that Geosource was involved in mapping equipment sales.

49. Offshore (Aug., 1983), p. 7.

50. Petroleum Economist (April, 1980), p. 177.

51. Sotsialisticheskaia industriia (Aug. 22, 1985), p. 1, translated in JPRS, UEN-86-007 p. 12.

52. Pravda (Feb. 12, 1984), cited in Soviet Geography, vol. 25, (Dec. 1984) p. 781.

53. Stroitel'naia gazeta (Nov. 23, 1983), cited in Soviet Geography, vol. 25 (Dec. 1984), p. 781.

54. EIU, no. 4 (1983), p. 97.

55. Canadian Department of National Defence, Soviet Offshore Oil and Gas Resources (unclassified), (Dec., 1984) p. 3.

56. This is suggested by a report in Sotsialisticheskaia industriia (Aug. 22, 1985), which says that two rigs (class and type are unidentified) were completed at BZGO. One weighed 3,500 tons, the other 7,000 tons. The first was taken to the 28th of April field, the other to the Mikhail Kaverochkin field, the field where it is known the "Shelf 2" is operating.

57. Petroleum Economist (April, 1984), p. 145-146.

58. Petroleum Economist (Aug. 1985), p. 293.

59. Sotsialisticheskaia industriia (Oct. 17, 1984), translated in JPRS, UEN-85-001 p. 103A. This will be dealt with in conjunction with the discussion on the Baltic and Barents seas.

60. Sovietskaia rossiia (Dec. 30, 1984), and Bakinskii rabochii (Jan. 30, 1985), cited in Soviet Geography, vol. 26 (March, 1985), p. 221.

61. Oil and Gas Journal (May 19, 1986), p. 44.

62. The legislation was passed by the Supreme Soviet in 1982 and was enacted as temporary legislation "permitting the licensing of Soviet companies to develop the seabed resources." See Offshore (June 5, 1982), p. 8.

63. Christian Science Monitor (Nov. 21, 1979), p. 7.

64. Offshore (June 20, 1979), p. 200.

65. Offshore (June 20, 1979), p. 200.

66. Offshore (June 20, 1979), p. 200.

67. Noroil (Aug., 1986), p. 111.

68. Information on Chukotsk from Sel'skaia zhizn' (Oct. 26. 1984), p. 1 translated in JPRS, UEN-85-001 p. 103B. Information on Kamchatka wells from Gazovaia promyshlenost, vol. 1, no. 7, (July, 1985), p. 11, translated in JPRS, UEN-85-018 p. 69.

69. Leslie Dienes and Theodore Shabad, The Soviet Energy System, (New York: John Wiley and Sons, 1979), p. 61.

70. James R. Lecky, J.R. Lee, "Siberian Oil", Siberian Resource Development (1979), pp. 113-123. See also Oil and Gas Journal (Dec. 22, 1975), p. 21.

71. P. Egyed, Western Participation in the Development of Siberian Energy Development: Case Studies, East-West Commercial Relations Series, No. 22, (Ottawa: Institute of Soviet and East European Studies, Carleton University, December 1983), p. 29.

72. P. Egyed (1983), makes note of this, ff p.38. Japanese concern over its trade position with the USSR vis-a-vis Western Europe is a theme that crops up quite often in Japan's dealings with the Soviet Union over Sakhalin.

73. Oil and Gas Journal, vol. 70, no. 49 (Dec. 1972), p. 46.

74. Oil and Gas Journal (Dec. 1972), p. 46.

75. East West Markets (Nov. 28, 1973), p. 3.

76. Soviet Business and Trade, vol.2, is. 21 (March 18, 1974), p. 1.

77. Soviet Business and Trade, vol. 2, is. 18 (Feb. 4, 1974), p. 1.

78. East West Markets (May 6, 1974), p. 6. East-West Trade Council Newsletter, vol. 2 no. 20 (Dec. 20, 1974), p. 5. P. Egyed (1983), asserts that at this time the agreement included an additional $55 million Japanese credit to finance equipment and consumer good purchases by the Soviet Union in Japan. This provision actually did not come about until late October 1974.

79. Oil and Gas Journal (June 3, 1974), p. 86.

80. The Economist (Oct. 22, 1977), p. 94.

81. East West Markets (Oct. 21, 1974), p. 3.

82. East West Markets (Oct. 21, 1974), p. 3.

83. Oil and Gas Journal (Nov. 3, 1975), p. 43.

84. East West Markets (Oct. 21, 1974), p. 3.

85. East West Markets (Oct. 21, 1974), p. 3.

86. East West Trade Council Newsletter, vol. 3, no. 2, (Jan. 31, 1975), p. 7.

87. Letter from E.E. Hansen (Gulf Oil Co.) to Carl McMillan, Director, Institute of Soviet and East European Studies at the East-West Project, Carleton University, Sept. 7, 1976.
88. See P. Egyed (1983), p. 48-49 and ff. p. 49.
89. V. Shushkov (USSR Deputy Minister of Foreign Trade), "Mutually Advantageous Cooperation Between the USSR and Japan", Soviet Foreign Trade (Moscow, May, 1975), p. 22-25.
90. World Oil (Aug. 15, 1977), p. 130.
91. East West Markets (April 19, 1976), p. 8.
92. Soviet Business and Trade, vol. 5, is. 23, (April 13, 1977), p. 2.
93. Soviet Business and Trade, vol. 5, is. 14, (Dec. 8, 1976), p. 1.
94. Soviet Business and Trade, vol. 5, is. 9, (Sept. 29, 1976), p. 2.
95. Soviet Business and Trade, vol. 5, is. 14, (Dec. 8, 1976), p. 1.
96. Soviet Business and Trade, vol. 5, is. 23, (April 13, 1977), p. 2.
97. Soviet Business and Trade (Aug. 17, 1977), p. 6. Information of Armco participation in the rigs equipping from OTA, (1982) p. 216. Levingston Shipyard Co. of Orange, Texas, U.S. went bankrupt in 1983. Conversations with its president that I had in August 1988 confirm all of the deals I have mentioned. The company is now simply known as Levingston.
98. Petroleum Economist (Sept., 1977), p. 372. Petroleum Economist (Nov., 1977), p. 449. Soviet Business and Trade (Oct. 21, 1977), p. 1.
99. New York Times (Oct. 13, 1977), p. 9.
100. East-West Markets (Aug. 7, 1978), p. 12.
101. Soviet Business and Trade (Aug. 7,1978), p. 12.
102. Moscow Narodny Press Bulletin (April 12, 1978), p. 10. Petroleum Economist (May 1978), p. 220. Soviet Business and Trade, vol. 6, is. 26, (May 24, 1978), p. 3. East West Markets (Sept. 18, 1978), p. 11.
103. Soviet Business and Trade, vol.6, is. 26, (May 24, 1978), p. 3.
104. Petroleum Economist (Jan., 1979), p. 36.
105. Platt's Oilgram News (June 8, 1979), p. 4.
106. Business Eastern Europe (July 13, 1979), p. 79.
107. Moscow Narodny Press Bulletin (Sept. 12, 1979), p. 15.

108. Soviet Business and Trade, vol. 8, is. 9, (Sept. 14, 1979), p. 2.
109. Soviet Business and Trade, vol. 8, is. 9 (Sept. 14, 1979), p. 3.
110. Petroleum Intelligence Weekly (Nov. 26, 1979), p. 12.
111. Soviet Business and Trade, vol. 8, is. 22, (March 12, 1980), p. 1.
112. Business Eastern Europe, vol. 9, no. 46, (Jan. 1980), p. 364.
113. Soviet Business and Trade, vol. 8, no. 23, (March 26, 1980), p. 1.
114. Business Eastern Europe (May 2, 1980), p. 138.
115. Business Eastern Europe (April 18, 1980), p. 128.
116. Soviet Business and Trade, vol. 8, is. 27, (May 21, 1980), p. 1.
117. Oil and Gas Journal (Nov. 10, 1980), p. 166.
118. Oil and Gas Journal (Nov. 10, 1980), p. 166.
119. Petroleum Economist (Oct., 1984), p. 387.
120. Soviet Business and Trade, vol. 10, is. 1 (June 21, 1981), p. 2. Soviet Business and Trade, vol. 10, is. 6, (Sept. 7, 1981), p. 2.
121. Business Eastern Europe (Jan. 15, 1982), p. 24.
122. Financial Times (Feb. 17, 1982), p. 6. Oil and Gas Journal (June 28, 1982), p. 47.
123. Financial Times (Feb 17, 1982), p. 6.
124. Soviet Business and Trade, vol. 11, is. 1, (June 29, 1982), p. 1.
125. This estimate was quoted in the Far Eastern Economic Review (July 23, 1982), p. 47.
126. Soviet Business and Trade, vol. 11, is. 4, (Aug. 7, 1982), p. 1. For information on the "Hakarya II" see the Petroleum Economist (Oct. 1984), p. 387.
127. Petroleum Economist (Sept. 1984), p. 356.
128. Petroleum Economist (Oct., 1984), p. 387.
129. Sotsialisticheskaia industriia (Oct. 17, 1984), translated in JPRS, UEN-86-001.
130. Oil and Gas Journal (Feb. 6, 1984), p. 92.
131. Petroleum Economist (June, 1986), p. 226.
132. Soviet Business and Trade (Dec. 23, 1984), p. 3.
133. Oil and Gas Journal (June 9, 1986), p. 29.
134. Petroleum Economist (June, 1986), p. 225.
135. Joseph P. Riva (1976), p. 492.

136. *Oil and Gas Journal* (April 26, 1976), p. 70.
137. *Financial Times of London* (Nov. 23, 1978), p. 5.
138. *Erdol und Kohle* (Feb., 1981), p. 52.
139. *Isvestiia* (April 15, 1984), p. 3.
140. *Erdol und Kohle* (Feb., 1981), p. 52.
141. *Petroleum Economist* (May, 1986), p. 188.
142. *Sovietskaia rossiia* (Feb., 1984), and *Isvestiia* (April 15, 1984), cited in *Soviet Geography*, vol. 25 (Dec., 1984), p. 781.
143. *Petroleum Economist* (Dec., 1984), p. 466, and *Oil and Gas Journal* (Dec. 3, 1984), p. 52.
144. *Sovietskaia rossiia* (Dec. 23, 1984), p. 6, translated in *JPRS*, UEN-85-006.
145. *Pravda* (Oct. 16, 1985), p. 5.
146. *Petroleum Economist* (June, 1986), p. 232.
147. *Petroleum Economist* (Sept., 1986), p. 351-352.
148. *Petroleum Economist* (Dec., 1984), p. 466.
149. James W. Clarke, *Petroleum Geology of the Offshore Soviet Arctic* (unpublished 1987), pp. 18-19. The reader should be aware of the ambiguity surrounding research on the petroleum geology of the Soviet Arctic. My work is focused on exploration and development activity in the Soviet offshore waters. Mr. Clarke's geological conclusions, therefore, should not be dismissed simply because there is a lack of Soviet activity in the Kara Sea. It is simply beyond the scope of this study to fully discuss Soviet Arctic petroleum geology. See also the discussion in the notes to Table 5.1 of this study.
150. *Ekonomicheskaia gazeta*, cited in *Oil and Gas Journal* (Aug. 25, 1986), p. 22.
151. The story of this blowout is interesting as an aside. A.A. Meyerhoff points out that the Bakhar-73 field blew in 1979, "...the entire rig and platform collapsed and caught fire. Water was used unsuccessfully to put out the fire. It broke out again repeatedly.... Finally, to extinguish the Bakhar blowout, a Soviet Army tank was brought to the location with a floating crane. Two 105-mm shells destroyed the valves and lateral pipes, so that the tongues of the fire that shot laterally outward from the well were shutoff. This left only a vertical flame which shot 80 meters upward. Press reports stated that, once the laterally shooting flames were extinguished, the vertical flame was brought under control 'by conventional means'. The well was finally under control on December 26, 1979 after nearly

eight months of 'blowing wild.' Officials boasted that the time to put out the fire was the shortest in the history of all offshore gas blowouts in the Soviet Union." See A.A. Meyerhoff (1980), p. 176.

152. Marshall Goldman (1980), p. 163.
153. Leonard LeBlanc, "Finland", Offshore (April, 1986), p. 124.
154. Vneshniaia torgovlia SSSR (1985), p. 172.
155. "Soviets to Press Drilling of SG-1 Superdeep Hole", Oil and Gas Journal (May 19, 1986), p. 23.
156. Oil and Gas Journal (June 14, 1976), p. 35
157. Financial Times (Oct. 22, 1986), p. 29.
158. Joseph P. Riva (1976), p. 496.
159. "Soviets Plan Arctic Work by 1982", Offshore (May, 1980), p. 162. See also, Noroil (Aug., 1984), p. 31.
160. Offshore (June 20, 1979), p. 199.
161. Offshore (May, 1980), p. 167.
162. Offshore (May, 1980), p. 168.
163. Offshore (June 20, 1979), p. 199.
164. Noroil (Aug. 1984), p. 31.
165. Noroil (Aug., 1984), p. 31.
166. Technical Information comes from the Petroleum Economist (April, 1982), p. 32. This source claims that the drillships are Russian-built, this is not the case and has not been corroborated any other source.
167. Noroil (Aug., 1984), p. 31.
168. "Energy Review", Economist Intelligence Unit, no. 3, (1983), p. 23.
169. Offshore (Aug., 1983), p. 17.
170. Oil and Gas Journal (April 7, 1986), p. 40.
171. Canadian Department of National Defence (Dec., 1984), p. 10.
172. Canadian Department of National Defence (Dec., 1984), p. 4 diagram, p. 6.
173. Information about the Kolguev strike is from Oil and Gas Journal (Feb. 14, 1983), p. 72. Speculation on the "Diorit's" activities there are my own.
174. Canadian Department of National Defence (Dec., 1984), p. 5.
175. "Energy Report", Economist Intelligence Unit, no. 4 (1983), p. 28.
176. Isvestiia (April 15, 1984), cited in Soviet Geography, vol. 25, (Dec., 1984), p. 782.

177. Canadian Department of National Defence (Dec., 1984), p. 8.

178. Canadian Department of National Defence (Dec., 1984), p. 4. For information on the "Chaivo" see Pravda (May 15, 1983), p. 1.

179. Leningradskaia pravda (March 5, 1985), p. 1, translated in JPRS, UEN-85-017.

180. Foreign Broadcast Information Service (FBIS), no. 8 (March 8, 1985), p. 3. The report stated that the value was over FM100 million, I have converted this to American dollars at the May 1987 rate of FM4.32 = US$1.00.

181. Offshore (April 1982), p. 19.

182. Offshore (June 20, 1982), p. 117.

183. A report in Oil and Gas Journal (Nov. 12, 1984), p. 64, suggests that the first of the Vyborg built semis the "Shelf 4" is in fact a modified Pacesetter design. This possibility cannot be ruled out, and indeed the whole fleet of Shelf rigs (seven in total now) may be nothing more than Pacesetter models. This wouldn't change the point too much however. The Soviet Union whether by reverse-engineering or improving on imported technology is nonetheless developing domestic capabilities to develop offshore rigging. One can use the Lada example to say that although the car is merely a copy of a 1972 Fiat, the Soviet Union has built a functional automobile industry based on that original import.

184. Petroleum Economist (July, 1985), pp. 254-255.

185. Oil and Gas Journal (Sept. 2, 1985), p. 30.

186. Oil and Gas Journal (Nov. 12, 1984), p. 64.

187. "In Search of Oil and Security: Soviet Interests in the Barents Sea", Noroil (Aug., 1986), p. 110.

188. "Viet Nam risks more offshore fusses with China", Oil and Gas Journal (July 11, 1988), p. 28.

189. Petroleum Economist (July, 1985), p. 255.

190. The names of the jack-ups are from Petroleum Economist (July, 1985), p. 255.

191. Oil and Gas Journal (Jan., 24, 1983), p. 32.

192. Petroleum Economist (July, 1985), p. 255.

193. This apparently is from a report out of Moscow. See Oil and Gas Journal Newsletter: USSR (Aug. 5, 1985), p. 6.

194. Petroleum Economist (July, 1985), p. 255.

195. Petroleum Economist (March 1985), p. 103.

196. Finnfacts, no. 10, (1984) cited in Moscow Narodny Press Bulletin (Nov. 21, 1984), p. 15.

197. Offshore (April, 1986), p. 124.
198. Oil and Gas Journal (April 15, 1985), p. 40.
199. This was also suggested in an article in Offshore (Nov., 1982), p. 8.
200. Offshore (Jan., 1984), p. 7.
201. See Offshore (June 5, 1983), p. 15, Petroleum Economist (June, 1983), p. 244. It should be noted that Soviet Business and Trade (July 9, 1984), p. 3, wrongly reported the Norwegian consortia as being named Conconor.
202. Noroil (Nov., 1984), p. 38.
203. See Petroleum Economist (April, 1984), p. 147.
204. This information came through Pomor-Nordic Trade, a joint Finnish-Norwegian company. The company was told by the Soviet Union that "...a joint license application by itself and a Norwegian oil company would stand a good chance of being accepted". See Petroleum Economist (June, 1985), p. 224.
205. Canadian Department of National Defence (Dec., 1984), p. 8.
206. Isvestiia (April 18, 1984), p. 6.
207. Oil and Gas Journal (Nov. 4, 1985), p. 59. This report is rather confusing in that it says that it was shipped from tundra fields in the far eastern sector of the Barents Sea. If true, it would indicate that while there is no production from the Barents Sea, the eastern sector is probably oil rich also.
208. Soviet Business and Trade, vol. 13, is. 17, (March 9, 1985), p. 1.
209. Future rig deployment is in Noroil (Aug., 1986), p. 110.
210. Globe and Mail (June 27, 1985), p. B5.
211. Business Eastern Europe (Sept. 13, 1985), p. 296.
212. Vneshniaia torgovlia SSSR (1985).
213. Oil and Gas Journal (Jan. 19, 1987), p. 29.
214. Oil and Gas Journal (Aug. 11, 1986), p. 34.
215. Sotsialisticheskaia industriia is quoted as the source on this, cited in Oil and Gas Journal (Aug. 11, 1986), p. 34.

5

The Move Offshore: An Assessment

At the outset of this study several questions were posed about the Soviet Union's practical ability to undertake exploration and development of its offshore waters. Summarized, these points are:

- *To what extent will problems in the onshore petroleum industry affect the decision to go offshore?*
- *Do estimates of offshore reserves suggest the region is worth exploring?*
- *What are the technological requirements for going offshore?*
- *Since the domestic industry is not capable at this time of fulfilling these requirements, how has the Soviet Union enlisted the assistance of foreign firms and governments?*
- *Given that foreign firms have been enlisted, how will the Soviet Union pay for this assistance?*

- *What are the security implications, and how has the Soviet Union dealt with the participation of foreign firms in the highly sensitive offshore waters?*

- *How are the effects of overriding political conflicts between East and West related to the use of and assistance from foreign partners?*

- *What are the systemic problems that inhibit the Soviet system's dealing with foreign firms, and what has the Soviet Union done to alleviate these problems?*

- *Finally, how has the Soviet Union dealt with or diffused the boundary disputes that have arisen over its exploration of the offshore waters?*

To what extent will problems in the onshore petroleum industry affect the decision to go offshore?

The extent to which problems with onshore petroleum production will make it necessary to consider offshore exploration and development is determined by two key factors.

First is the declining reserves-to-production ratio. As shown in Chapter 2, the Soviet Union has had a declining reserves-to-production ratio in petroleum of -3.9% over the past decade. Exploratory drilling onshore has not been as successful as hoped, and the technical support to insure its increase was harshly criticized by Gorbachev in his visit to the Tiumen' district in the fall of 1985.[1]

It is clear to Soviet planners that offshore petroleum will not compensate for the decline in the large fields, such as Samotlor or Fedorovo in the Tiumen' district. Many Soviet geologists are pessimistic about the ability of West Siberia to add significantly to explored reserves. Further, it is now realized that traditional means of increasing oil production, i.e., at the expense of exploratory drilling, will lead to a significant drop in oil production in the next two decades generally, and in West Siberia specifically.[2]

The Gorbachev administration has, therefore, launched what seems to be a more comprehensive strategy for the petroleum industry. While this strategy includes such approaches as the formation of a super-ministry in the whole energy sector, it also includes an increase in exploration activity in areas of large potential reserves. This has included both East Siberia and the offshore waters. Exploration and development of the offshore regions have, therefore, been part of the general strategy to insure long-term petroleum production.

The second determining factor is the importance of petroleum production to export earnings. As the quality of Soviet products continues to lag behind similar products in Western markets, petroleum production and future production (accessible reserves) are extremely important. Offshore petroleum has a somewhat contradictory role in this. On the one hand, offshore petroleum's eventual contribution to reserves and production will increase petroleum for export to hard currency markets. More immediately, on the other hand, the extent to which hard currency is required to explore and develop the offshore regions, to date, is enormous. It would seem, however, the present leadership is concerned with exploring and developing the offshore waters, as evidenced in the 27th Party Congress's declaration to increase the development of the petroleum reserves on the continental shelf.[3] This is part of a two-pronged approach by Gorbachev. He is also trying to reduce the share of fuels generally in Soviet exports, but this is largely a function of his ability to reform the economy and stimulate high-quality exportable products.[4] It is necessarily, therefore, a long-term goal which goes far beyond the complexity of offshore exploration and development.

Do estimates of offshore reserves suggest that the region is worth exploring?

The areas with the best petroleum reserves are the areas that will receive the most attention in the Soviet Union's exploration and development plans. As shown in Table 5.1, the total estimated reserves of the offshore waters are between 98 and 151 bbl. These figures, however, are quite tenuous given the rough estimates for the Arctic waters' reserves between 40 and 80 bbl. There are really only two regions of proven reserves, the Caspian Sea and the Sea of Okhotsk. The reasons for this are as follows.

In the case of the Caspian Sea, a moderate climate and proximity to a traditional oil-producing province afforded easier access and, hence, allowed for better seismological surveys of the basin. Further, the advantage of the shallow Apsheron Sill made exploration activity much easier. When the Soviet Union was ready to venture into deeper waters it was able to undertake seismological surveys well in advance of exploratory drilling, allowing for much more success in the wells it actually drilled.

Table 5.1
USSR Offshore Petroleum: Regions and Key Fields, Ultimate and Explored Reserves, Current Production

Region and Key Fields	Ultimate Regional Reserves (bbl)	Explored Reserves (bbl)	Current Production or Test-pay* (bd)
Caspian Sea	30-35	4.0	190,000
Adrella-more	n.a.	n.a.	n.a.
Alytai-more	n.a.	n.a.	n.a.
28th of April	n.a.	1.0	90,000
Andreev bank	n.a.	n.a.	n.a.
Bakhar field	n.a.	n.a.	2,770*
26 Baku Commissars	n.a.	1.0**	2,373*
Barinov bank	n.a.	n.a.	n.a.
Bulla island	n.a.	n.a.	n.a.
Cheleken	n.a.	n.a.	n.a.
Garasu	n.a.	n.a.	1,825*
Gubkin bank	n.a.	n.a.	n.a.
Kaverochkin	n.a.	n.a.	n.a.
LAM bank	n.a.	n.a.	n.a.
Livanov East bank	n.a.	n.a.	n.a.***
Neftianye Kamni	1.2	1.2	60,000
Promezhutochnaia	n.a.	n.a.	1,460
Rakushechnoe-more	n.a.	n.a.	n.a.
Zapadnoe-Okarern	n.a.	n.a.	n.a.
Zhdanov bank	n.a.	n.a.	n.a.

Continued on next page

Table 5.1 continued

Region and Key Fields	Ultimate Regional Reserves (bbl)	Explored Reserves (bbl)	Current Production or Test-pay* (bd)
Black and Azov seas	n.a.	n.a.	n.a.
Karkinit Deep	region possibly oil and gas rich		
Sea of Okhotsk	28-36	1.0	0.0
Chaivo field	n.a.	0.6	2,050*
Odoptu field	0.7	0.4	7,000*
Dagi field****	n.a.	n.a.	n.a.
Baltic Sea	n.a.	n.a.	n.a.
27-oi S"ezd KPSS	n.a.	n.a.	unknown*
Arctic Waters	40-80	n.a.	n.a.
Barents Sea*****	n.a.	n.a.	n.a.
Pechora Sea	n.a.	n.a.	n.a.
Kara Sea*****	n.a.	n.a.	n.a.

Notes: Reserve estimates for the Caspian Sea are from E.N. Tiratsoo, *Oilfields of the World*, Scientific Press (1976). Here it states that the inland sea reserves are roughly 39 bbl. The *Petroleum Economist* (May, 1977), p. 191 estimated the Caspian reserves to be around 30-35 bbl. Promezhutochnaia estimate from *Petroleum Economist* (March 1988), p. 96.

Reserve estimates for the Sea of Okhotsk are from a Soviet source cited by Junnosuke Ofusa "Japan and Soviets Agree on Joint Oil and Gas Project," *New York Times* (November 25, 1976), p. 41. This estimate is corroberated by E.O. Egleograh and E.J. Peters, "The Importance of the Soviet North in the Total Soviet and World Energy Supply System," *Conference on Energy in the USSR and Eastern Europe*, University of Alberta (May 27-28, 1977), p. 21. Similar estimates can be found in *Oil and Gas Journal* (June 10, 1974), p. 95.

Estimates for the Arctic are extremely tenuous. The cited amount is based upon my conversations with geo-

Continued on next page

Table 5.1 Continued

logists at the Department of Energy Mines and Resources, Ottawa, Canada in July 1987.

* Test-pay figure.

** Reports claim explored reserves of over 1.0 bbls.

*** Three wells have been drilled, test-pays unknown.

**** Dagi is an onshore Sakhalin field with apparent offshore extensions. According to A.A. Meyerhoff (1980), p. 174, a successful wildcat well was drilled east of Dagi in September 1980.

***** James Clarke, personal communication, April 15, 1988. Clarke, of the *U.S. Geological Survey*, notes that the Barents Sea had petroleum reserves of around 8 bbl oil, while the Kara Sea has reserves of about 6 bbl oil and 164 trillion cubic feet of gas. The Kara estimates are easier to accept since there seems to be some consensus. However, Clarke's estimates on the Barents are questionable and have been disputed in a number of sources. The Barents reserves may be much higher, by twice or three times.

The Sea of Okhotsk has had a reasonable assessment of its reserve size due to the Soviet Union's cooperation with Japan. While the Sakhalin offshore petroleum project was not a success from the point of view of production, it was a success from the point of view of the amount of seismological work and exploratory drilling that went on before the relationship became spoiled by American sanctions and falling world oil prices. In this sense the Soviet Union achieved a major goal of assessment of an offshore region.

The other two areas of more moderate temperatures, compared to the Sea of Okhotsk and the Arctic, are the Baltic Sea and the Black and Azov seas. Here, while the potential reserves are not known, they are believed to be low, at least in petroleum. The only criterion to justify costly investment in offshore activity would be to satisfy local demand. These areas, then, will probably not warrant significant expenditures in the coming years.

The area which will be the most heavily explored in the next decade, however, will be the Barents Sea, specifically the western portion of the Barents Sea. The only information known is that Soviet geologists have estimated the Barents Sea to be petroleum-rich. The Kara Sea may also be petroleum-rich below 6,000 meters. The other seas of the Arctic are not thought to be as petroleum-rich as the Barents, but far too little is really known of these basins to afford educated speculation.

It would seem the Barents Sea is at present a place of exploration activity due to its proximity to Norway (explaining Norwegian interest in the zone with the Soviet Union), and because part of the sea is ice-free. The extent to which the other seas will become areas of concern will be dependent upon how well the Soviet Union is able to overcome problems in ice-bound waters. One should bear in mind that the seismological estimates of all other regions, save the American section of the Bering Sea, is largely based upon onshore seismological work and is not, therefore, entirely dependable. As the required technology is gained, then the Laptev, East Siberian, and Chukotsk seas may prove to be more petroleum-rich than currently thought.

What are the technological requirements for going offshore?

For the Soviet Union the issue is really one of how exploration and development of their offshore reserves differs from onshore exploration and development.

The first problem the Soviet Union encountered was that of exploratory drilling in the Caspian Sea. Until 1974, offshore exploration and production had been treated much the same as onshore production by dropping large metal trestles at prospective sites. These trestles, however, were immobile, severely constraining the development of offshore exploration. The first jack-up rig, bought from the Dutch (see Table 5.2) was a significant purchase. It allowed the Soviet Union some mobility of exploration in offshore waters and made an essential difference from the onshore industry. To move rigs from site to site as more petroleum was recorded on seismological surveys would soon become extremely expensive. Therefore, the mobile jack-ups (with retractable legs) were quite a breakthrough.

Jack-ups, however, are only useful if seismological surveys are fairly complete; if the shelf where the deposit is discovered is at a fairly shallow depth (around 60 meters); and if there is no extensive exploratory drilling necessary. The most important consideration is the water depth. The Soviet Union discovered in both the Caspian Sea and Sea of Okhotsk that it would only be possible to explore the whole shelf in deeper water. For this task semisubmersibles were required. It is here the offshore industry diverges drastically from its onshore counterpart. In order to undertake exploratory drilling and production in waters greater than 60 meters, semisubmersibles are essential. The technological requirements to build and operate one are enormous, as the Soviet Union discovered. While it was possible to copy the Dutch-built jack-up purchased in 1967, the Soviet Union has had only limited success in developing its own semisubmersible units. Indeed, some observers believe that the Soviet Union simply purchased outright from Finland the seven Shelf units it has operating in its offshore waters.

Another complication in exploring and developing the offshore regions is that many of the potential fields on the Soviet shelf are in ice-bound waters. Unlike the experience of the onshore industry, ice sheer and buckling is much more drastic (i.e., costly) in deep waters than on land. As a result, the Soviet Union was forced to import and operate ice-resistant rigging in the Sea of Okhotsk and some sectors of the Arctic. This is an import dependency the Soviet Union has been unable to overcome to date.

Finally, in order to quicken the pace of exploration and to increase the certainty of reserves estimates in the offshore waters, sophisticated drillships are required. These drillships must be designed for very deep waters (beyond 500 meters) and must also have special computer-aided propellers to compensate for wave action on the drillship. The drillships are also more delicate than a jack-up or a semisubmersible in that the drill rods do not have as much weight to rely on at extremely low depths. The rods tend to be more brittle. Correct propeller operation is essential to compensate for wave action on the vessel.

These are the salient technological requirements facing the Soviet Union in the offshore petroleum industry. If one adds to this list all the support requirements, such as ice-resistant sleeves for Arctic jack-ups, manned diving bells for arctic conditions, special pipe-laying barges to onshore sites and blowout equipment specific to offshore accidents, then both the requirements and the cost of offshore technology become staggering.

One of the most important points stressed in this study is the nearly complete lack of domestic technology to meet any of these requirements. The Soviet Union, therefore, has enlisted a number of foreign firms to assist in its acquisition and operation of the required technology.

Since the domestic industry is not capable at this time of fulfilling these requirements, how has the Soviet Union enlisted the assistance of foreign firms and governments?

The real question in all of this is accessibility, which to the Soviet Union means foreign (Western) technology. The Soviet Union has a nearly complete dependence upon Western equipment and expertise in the exploration and development of the Soviet Union's offshore waters. As was pointed out in Chapter 4, the only area with any predominance of Soviet expertise was the Caspian Sea exploration and development in its early stages. Once it became necessary to move off the Apsheron Sill into the deep waters of the Caspian Sea it became essential to enlist foreign help.

The Soviet Union has enlisted foreign help in two ways. The first is with the outright purchase of foreign technology. The second is with the participation and cooperation of foreign firms.

Predictably, the Soviet Union began to simply purchase foreign equipment in the early stages of offshore exploration and development in the Caspian Sea. The purchase of the Dutch-built jack-up, the *Khazar* was the first example of this. It appears the Soviet Union was able to reverse-engineer this rig quite successfully. By 1978 there were five Soviet-built jack-ups operating in the Caspian Sea. In the Sea of Okhotsk, however, the Soviets were not quite as successful. The technological constraints of ice-filled waters complicated the construction and use of jack-ups. The Soviet Union was forced to allow the *Interocean I*, operated by the Japanese, to explore the waters off Sakhalin. When

Table 5.2: USSR Imports of Offshore Technology, 1967 to 1987

Year	Company Name	Country	Description	Value	Destination
1967	IHC	Netherlands	Delivered the jack-up rig "Khazar".	n.a.	Caspian Sea
1971	IHC	Netherlands	Delivered sea going pipe-laying barge "Suleiman Yeyirov".	n.a.	Caspian Sea
1975	Serete	France	Delivered a floating drilling platform.	n.a.	Caspian Sea
1975	Gulf Oil Co.	U.S.	Committed one of its advanced seismic exploration vessels.	n.a.	Sea of Okhotsk
1976	Rauma-Repola Oy	Finland	Constructed hull for semisubmersible "Kaspmorneft'".	$15 million	Caspian Sea
1976	Armco	U.S.	Supplied equipment along with Rauma Repola for the "Kaspmorneft'".	n.a.	Caspian Sea
1976	Blohm & Voss AG	West Germany	Delivered a self-propelled crane to position offshore rigs.	$40 million	Caspian Sea
1976	Sanwa Kizai Co.	Japan	Delivered 14 augers used for pile driving.	n.a.	Sea of Okhotsk
1976	National Supply Co. (Armco subs.)	U.S.	Provided most of the equipment for a semi-submersible rig being constructed by Rauma-Repola Oy.	$25 million	Caspian Sea
1976	Armco	U.S.	Provided license for the production of semisubmersible rigs in the Soviet Union.	$25 million	Caspian Sea (Astrakhan yard)

1976	Compagnie Generale Geophysique	France	Leased the geophysical research vessel "Orion Arctic".	n.a.	Sea of Okhotsk
1976	SODECO	Japan	Leased two geophysical reserch vessels the "Ranger Maru" and the "Hercules".	n.a.	Sea of Okhotsk
1976	Interocean	Japan	Leased a jack-up rig, "Interocean I".	n.a.	Sea of Okhotsk
1976	Suitomo Heavy Industries Ltd.	Japan	Built the Interocean drilling platform	n.a.	Sea of Okhotsk
1976	Gulf Oil Co.	U.S.	Provided the Soviet Union with 20 specialists to assist in the operation of the Interocean rig.	n.a.	Sea of Okhotsk
1976	(firm unknown)	France	Delivered digital seismographic recorders, magnometers, gravity meters, cables and hydrophones. The French also provided training.	$14 million	Sea of Okhotsk
1976	Mitsubishi Corp.	Japan	Installed the French equipment on two Soviet rigs.	$2.5 million	Sea of Okhotsk
1977	Kyoho Tsusho Kaisha Ltd.	Japan	Sold 4 20-ton floating cranes.	Countertrade deal for hard alloy drills from Stankoimport.	Sea of Okhotsk.
1977	Fukoshima Ltd.	Japan	Sold 4 floating cranes to Soviet Union.	$10 million	Sea of Okhotsk
1977	Levingston Shipyard Co.	U.S.	Provided design for Mitsui jack-up.	n.a.	Sea of Okhotsk.

Continued on next page

Table 5.2 Continued

Year	Company Name	Country	Description	Value	Destination
1977	Mitsui Engineering and Shipbuilding Co. Ltd.	Japan	Soviet Union ordered a jack-up from Mitsui.	n.a.	Sea of Okhotsk
1977	Otis Engineering Corp.	U.S.	Provided on-board equipment for Mitsui rig.	n.a.	Sea of Okhotsk
1977	Armco	U.S.	Provided equipment for Mitsui rig.	n.a.	Sea of Okhotsk
1977	Dolphin International Ltd.	U.S.	Leased the jack-up "Borgsten-Dolphin" to replace "Interocean I".	n.a.	Sea of Okhotsk
1978	Armco	U.S.	Licensed parts to Rauma-Repola Oy for semisubmersible being built for the Soviet Union.	n.a.	(unknown probably Arctic)
1978	Ulstein Hatlo A/S	Norway	Sold two supply vessels.	n.a.	Sea of Okhotsk
1978	Laivatellises Oy	Finland	Sold two supply vessels.	n.a.	Sea of Okhotsk
1978	Skaugen Offshore Supply Ships	Norway	Provided the contract and participated with Laivatellises Oy on the research vessels construction.	n.a.	Sea of Okhotsk
1978	Geosource	U.S.	Sold two Command II field processing systems for geophysical mapping.	n.a.	Caspian Sea

1978	Foremost Industries Ltd.	Canada	Supplied 24 tracked geophysical survey vehicles to carry Geosource equipment.	n.a.	Caspian Sea.
1978	Rotterdam Shipyards	Netherlands	Were contracted to build a self elevating offshore drilling rig.	$49 million	Baltic Sea
1978	Rhein-Schelde-Verolme	Netherlands	Contracted to build a drilling plat-form, to be delivered in the mid-1980s.	n.a.	Baltic Sea
1978	Pusnes Mek-aniske Verk-sted A/S	Norway	Licensed cold-weather mooring technology.	n.a.	Barents Sea
1978	Elf Aqui-taine	France	Agreed to a general financing and technology cooperation agreement in the Barents.	n.a.	Barents Sea
1979	Blohm & Voss AG	West Germany	Contributed to the construction of of the Astrakhan yard.	n.a.	Caspian Sea
1979	Mitsui Ocean Development Co.	Japan	Delivered a self-lifting drill rig.	$36 million	Sea of Okhotsk
1979	Mitsui Ocean Development Co.	Japan	Provided a class III three-legged jack-up rig--the "Hakarya II".	$35 million	Sea of Okhotsk
1979	Levingston Ship-yard Co.	U.S.	Sold the "Harkarya II" design to Mitsui.	n.a.	Sea of Okhotsk
1979	Armco Inc.	U.S.	Through its subsidiary National Supply Division provided equipment for the "Hakarya II".	n.a.	Sea of Okhotsk

Continued on next page

Table 5.2 continued

Year	Company Name	Country	Description	Value	Destination
1979	N.L. Petroleum Services	U.S.	Provided two blow-out preventor stacks to the "Hakarya II".	$525 thousand	Sea of Okhotsk
1979	Baker International	U.S.	Provided testing equipment for the "Harkarya II".	n.a.	Sea of Okhotsk
1979	Kongsberg Vaapen Fabrikk A/S	Norway	Provided three dynamic positioning systems for three drillships being constructed by Rauma-Repola Oy.	n.a.	Barents Sea
1979	Simard A/S	Norway	Provided dynamic positioning system for Armco-licensed semisubmersible.	n.a.	Caspian Sea
1979	Lynes International Inc.	U.S.	Sold 11 strings of drill stem-testing equipment.	$3.8 million	(unknown)
1979	National Supply Division	U.S.	Sold drilling equipment and a jacking system for a jack-up rig.	$4 million	(unknown)
1980	ETPM	France	Assisted in developing exploration vehicles.	$59 million	Caspian Sea
1980	UIE and ETPM	France	Assisted in construction of Baku yard at Primorsk.	n.a.	Caspian Sea
1980	Oceanic Contractors Inc. (McDermott Inc. subs.)	U.S.	Undertook a feasibilty study of permanent fixed platform drilling.	$1.5 million	Sea of Okhotsk
1980	UIE	France	Assisted in developing exploration vehicles.	$59 million	Caspian Sea

1980	McDermott Inc.	U.K.	Major equipment deal (presumably for ice-resistant rigging).	$450 million	Sea of Okhotsk
1980	Mitsui Ocean Development Co.	Japan	Constructed and delivered the jack-up "Okha".	n.a.	Sea of Okhotsk
1980	Armco	U.S.	Supplied equipment to Mitsui the "Okha".	n.a.	Sea of Okhotsk
1980	Offshore Supply Association	U.K.	Operated two supply vessels in the Chaivo bay area.	n.a.	Sea of Okhotsk
1980	(unknown)	Netherlands	Jack-up rig, delivered to Gdansk, (Poland) yard in 1980, then sent to the Soviet Union in 1984.	n.a.	Baltic Sea
1981	Rauma-Repola Oy	Finland	Soviet Union placed an order for three dynamically positioned drillships, the "Valentin Shashin", "Mikhail Mirchink" and the "Viktor Muravienko".	$225 million	Barents Sea
1981	IHC-Gusto Pelican	Netherlands	Licensed design of "Valentin Shushin" to Rauma-Repola Oy (Finland).	n.a.	Barents Sea
1981	RSV Gusto Engineering	Netherlands	Undertook much of the subcontracting work for Rauma-Repola Oy on the three drillships.	n.a.	Barents Sea
1982	Keppel Shipyards	Singapore	Repaired the "Mikhail Mirchink" after its use off Vietnam.	$2.8 million	(unknown)

Continued on next page

Table 5.2 Continued

Year	Company Name	Country	Description	Value	Destination
1983	Far East-Levingston Shipbuilding Co.	Singapore (Keppel Shipyard Co. subs.)	Sold jack-up rig and crane vessel.	n.a.	(unknown)
1983	Blohm and Voss AG	West Germany	Sold a 25,000 ton launch barge to launch jackets for water depths up to 700 ft.	n.a.	Caspian Sea
1983	Comexindustrie	France	Provided manned diving bells "SM-170" and "Moana". The French are also providing training.	n.a.	Barents Sea
1983	(unknown)	Finland	Delivery of the semisubmersible "Vyborg I", possibly Soviet built.	n.a.	Barents Sea
1983	Rauma-Repola Oy	Finland	Delivery of four Pacesetter semisubmersible rigs. Probably the first four Shelf units.	n.a.	Barents Sea Caspian Sea Sea of Okhotsk
1983	McDermott Inc.	U.K.	Sold preliminary engineering equipment and conceptual designs for ice-resistant rigging production.	n.a.	Sea of Okhotsk Barents Sea
1984	Kone	Finland	Delivery of a catamaran crane vessel the "Titan-I" for hoisting and installing for offshore oil fields (1 of 4).	n.a.	Barents Sea Vyborg Yard
1984	ENI	Italy	Provided deep-water pipeline technolgy.	n.a.	Barents Sea

1984	Sirtec	Italy	To provide equipment for Arctic rigs and underwater pipelines at low depths.	n.a.	Barents Sea
1984	IRI	Italy	Provided deep-water pipeline technology.	n.a.	Barents Sea
1985	Seaforth Maritime	Scotland	Supplied advanced saturation technology for the "Shelf 4".	$10 million	Barents Sea
1985	Rauma-Repola Oy	Finland	Delivered two Arctic-class jack-ups "Kolskaia" and "Sakhalinskaia".	$180 million	Barents Sea Sea of Okhotsk
1985	Parkano Engin-eering Works	Finland	Constructed the 37 elevating units on the "Kolskaia" and "Sakhalinskaia".	n.a.	Barents Sea Sea of Okhotsk
1985	Saga Petroleum	Finland	Arctic technology (type unknown).	n.a.	Arctic (Area unspecified)
1985	Valmat	Finland	Arctic production platforms, modules and topsides.	n.a.	Arctic
1985	Foster Wheeler Petroleum Development	U.K.	Undertook cooperation deal with Valmat to assist in the development of arctic platforms, modules and topsides.	n.a.	Arctic
1985	Rauma-Repola Oy	Finland	Promised to produce and deliver a floating production unit and a mobile arctic drilling and production rig.	n.a.	Barents Sea
1985	Brown and Root Co.	U.K.	Joint-ventured with Rauma-Repola Oy on floating production unit for the Arctic.	n.a.	Barents Sea

Continued on next page

Table 5.2 Continued

Year	Company Name	Country	Description	Value	Destination
1985	Rauma-Repola Oy	Finland	Constructed the Soviet-designed research vessel "Dimitrii Nalivkin" and is to have delivered one more of Soviet design in 1986 or 1987.	$23.1 million	Barents Sea
1986	Nobara Sociedade de Mineracao Comercio & Industria	Brazil	To provide Soviet Union with 10 offshore oil drilling platforms.	Countertrade deal with Petrobas (Brazil) for petroleum.	Barents Sea Caspian Sea
1986	Versatile Corp.	Canada	Will provide expertise in the construction of icebreakers.	Versatile will receive technical information on Arctic geology.	Barents Sea Kara Sea Laptev Sea Arctic Ocean
1987	Neste	Finland	To provide a geotechnical research ship, an offshore drilling and transport rig, two accommodation ships, an underwater exploration and rescue vessel, two Arctic-class jack-up legs and a jack-up unit.	Countertrade deal in which Neste buys 11.5 million tons of oil and reexports 2 million tons.	Arctic

Source: See Chapter 4.

the *Interocean I* ran aground it was replaced by the American-built rig, the *Borgsten-Dolphin*. It is an indication the Soviet Union really did need to cooperate with Western firms (in this case, Japan) to acquire the necessary technology. In 1979, instead of constructing their own rig the Soviet Union was forced to lease the *Hakarya II* (fully equipped with American-licensed parts) to begin exploration off Sakhalin's northeast coast. This rig was not successfully copied, if the Soviets even tried. The *Okha* was the next rig to operate in the Sea of Okhotsk and was sold to the Soviet Union by the Japanese in 1980.

When it became evident that simply purchasing Western equipment was not enough and that Western expertise would be necessary, the Soviet Union began to engage in cooperative efforts with other nations. The most important case was that of Japan. A formal agreement on exploration and development in the Sea of Okhotsk was reached in 1975. Both Finland and, to a lesser extent, Norway have been involved in discussions on development of specific regions of interest to all three countries. In the case of Finland more general deals have been negotiated, such as the countertrade deal with Neste Oy (see Table 5.2) in 1987.

Two key objectives that have distinguished the Soviet Union's involvement with foreign firms. One minimizes the Soviet Union's financial obligations to foreign partners; the other minimizes the number of foreigners (engineers, observers, etc.) working in Soviet waters.

First, cooperation with Japan showed the Soviet Union was trying to strike a deal where it could attract as much foreign technology as possible, but with minimal foreign influence. Minimal foreign influence meant that the Soviet Union could not be held financially responsible for a lack of hydrocarbon discoveries. In the agreement with Japan, the Soviet Union fought hard and won the right not to repay a $100 million loan from Japan should hydrocarbons not be discovered. Minimal foreign influence also meant that few if any foreign engineers would be allowed in the Soviet zones of exploration. This leads to the second feature.

The Soviet Union has always tried to restrict the number of foreign participants, usually engineers, actually operating in its offshore waters. Indeed, negotiations with Japan were almost halted in 1975 when Japan refused to back down on its desire to have observers aboard the survey ships in the

Sea of Okhotsk in the Soviet zone. The second condition represented a turning point because it was conceded.

In the western Barents the Soviet Union seemed willing to allow the operation of either Norwegian or Finnish vessels. Cooperation with Norway, via the consortium BOCONOR, including exchanges of technical information, ice-resistant platforms, and sub-sea production units, have been complemented by the Soviet Union's considering granting licenses to Norwegian firms for the exploration of the west Barents. This consideration is surely on the heels of the early 1982 joint-venture legislation that allowed joint ventures with foreign firms in Soviet waters in order to accelerate the pace of offshore development.

Given that foreign firms have been enlisted, how will the Soviet Union pay for this assistance?

Table 5.2 is revealing in a number of ways. First, comparing it to Table 5.3 reveals that all of the mobile rigs operating on the Soviet shelf are either in whole or predominantly in part made from Western technology. Second, the actual expenditure on foreign technology has only been an estimated $9.4 billion.[5] Although this is a fairly small portion of the total hard currency revenues gained from petroleum sales to the West in the same period, one year is particularly revealing.

As shown in Table 5.2, 1976 is a year in which 6 of 13 imports are known in value. During that year it is estimated approximately $526.5 million was spent on offshore equipment. In that same year total hard currency earnings from petroleum equalled $5.2 billion. Over 10% of the hard currency earned in 1976 was offset by expenditures on offshore equipment. In 1976 offshore production was only 2.1% of all petroleum production. It substantially declined after that year and fell to around 1.5% [6] The initial capital outlay appears quite high for the return, at least in terms of production.

Table 5.3
Soviet Mobile Offshore Rigs and Research Vessels Currently Operating in Offshore Waters

Region	Year of Introduction and Vessel Name	Type of Vessel
Caspian Sea	1974 Apsheron	Jack-up
	1974 Khazar	Jack-up
	1974 60 let Azerbaidzhana	Jack-up
	1975 Baku	Jack-up
	1978 60 let Oktiabria	Jack-up
	1981 Kaspmorneft'	Semisubmersible
	1981 Shelf 1	Semisubmersible
	1983 Shelf 2	Semisubmersible
	1985 40 let Pobedy	Jack-up
	1986 Shelf 3	Semisubmersible
Sea of Okhotsk	1977 Okha	Jack-up
	1977 Ekhabi	Jack-up
	1983 Mikhail Mirchink*	Drillship
	1985 Sakhalinskaia	Jack-up
	1987 Shelf 5**	Semisubmersible
Black and Azov seas	1981 Sivash	Jack-up
	1986 Golitsynoe 1-5	Semisubmersible
Barents Sea	1978 Professor Shtokman	Research Vessel
	1982 Valentin Shashin	Drillship
	1982 Viktor Muravienko***	Drillship
	1984 Chaivo	Drillship
	1984 Akademik Kurchatov	Research Vessel
	1984 AkademikGolitsyn	Research Vessel
	1984 Spurt	Research Vessel
	1985 Kolskaia	Jack-up
	1985 Shelf 4	Semisubmersible
	1986 Shelf 6**	Semisubmersible
	1986 Dimitrii Nalivkin	Research Vessel
	1987 Diorit	Drillship
	1987 Shelf 7	Semisubmersible
Baltic Sea	1986 (unknown)	Jack-up

Continued on next page

Table 5.3 continued

Notes: * The *Mikhail Mirchink* is either operating off the Kamchatka peninsula near the American border or, as some reports state, was damaged off the coast of Vietnam some time in 1984 and is being towed to a Singapore shipyard for repairs.
** The present locations of the *Shelf 5* is unconfirmed. This is the author's best guess. The *Shelf 6* was originally deployed in the Barents Sea but in May 1988 was sent to Danhung in the South China Sea off the coast of Vietnam.
*** The Viktor Muravienko was reported to be moving to the Kara Sea to begin drilling. See *Petroleum Economist* (January 1988), p. 30.

Sources: See Chapter 4. A similar table that lists some of the above drillships as well as planned rigs is given in H.O. Bergsen, A. Moe and W. Ostreng, "In Search of Oil and Security in the Barents Sea," *Noroil* (August 1986), p. 110.

While data is sparse for later years, it is quite possible the proportion of expenditures on foreign equipment to total import earnings has in fact increased. In 1976, the highest price paid for an item was $40 million. By 1980, this had increased to $450 million. The lowest known purchase price in that year was $225 million. The Soviet Union has increased the number of rigs bought from foreign firms also since 1980, as shown in Table 5.2. In 1986 alone, the Soviet Union seems to have purchased over ten drilling platforms and at least three partial or whole jack-ups costing around $200 million each. When this is viewed in light of declining revenues from petroleum exports since 1984, it would suggest an increase in the ratio of the value of offshore equipment imports to import earnings from petroleum.

There is every possibility the value of the ratio of offshore equipment imports to petroleum export earnings will increase if current trends in petroleum export earnings continue. The cost of ice-resistant rigging is enormous, as evidenced by the McDermott Inc. (U.K.) sale of ice-resistant rigging for use in the Sea of Okhotsk at $450 million. As this type of equipment will become more and more essential, due to increased exploration of the Arctic region, one can expect the costs to increase at a rapid pace for the next decade.

The ability of the Soviet Union to pay for further acquisitions of offshore technology is a function of how much confidence the Soviet Union has in the technology's ability to pay for itself and eventually significantly contribute to overall petroleum production in the Soviet Union. If energy planners are convinced of the reserve capacity and accessibility of offshore oil, then imports can be paid for by dedicating a certain amount of hard currency earnings from present petroleum exports to paying for these imports. The hope would be that eventually the future yield from the offshore fields would compensate for expenditures on equipment today. However, this is a rather risky option, especially since some of the major costs are not for production equipment but for exploration equipment. It is this equipment that is necessary for planners to assess the capability of offshore fields to yield enough oil to pay for itself. In the short term the Soviet Union could do several other things to pay for more offshore equipment imports. The transportation and food-processing division of agriculture currently runs at 20% to 30% waste.[7] Improvement here would help reduce grain imports of up to $4 billion. In the more immediate future the traditional method of increasing gold sales could provide some hard currency.

Probably the method the Soviet Union will stress is countertrade. Indeed, the last three deals the Soviet Union has negotiated have been countertrade deals as evidenced in Table 5.2. However, problems with countertrade deals are well-known. For many Western firms it is difficult to sell Soviet goods on Western markets, and the Soviet partner tends to undervalue the Western input. Another option that may be considered by Soviet planners is the leasing of Western equipment. Such an option would take advantage of the Soviet Union's excellent credit rating. This, and the fact that its realizable assets in the West exceed its gross debt of around $23 billion by a substantial amount, puts the Soviet Union in a good position to lease equipment.[8]

Finally the foreign trade reforms under Gorbachev have had the effect of increasing enterprise autonomy to deal with foreign firms. Especially important is the Soviet enterprise's sole rights to its hard currency earnings and special drawing rights from Vneshtorgbank for imports that it can guarantee repayment in hard currency of in four years time.[9] This is part of the effort to encourage joint ventures and can be viewed as one of the ways in which to reduce the hard currency requirements from oil export earnings. The fact that joint ventures will not be subject to plan targets may be attractive to Western partners.[10]

What are the security implications and how has the Soviet Union dealt with the participation of foreign firms in the highly sensitive offshore waters?

Security considerations are vital and must be viewed as a major inhibiting factor to the involvement of foreign firms. There is the general consideration that petroleum is classified as a strategic resource and is, therefore, subject to a considerable amount of secrecy. Beyond this, however, the offshore waters present a general security problem due to their location. The most important of these waters is the Barents Sea.

The Soviet Union's largest naval base is on the Kola peninsula. The Barents Sea is the only access route. The waters of the western Barents, the site where most of the exploration of both Norwegian and Soviet vessels is taking place, is the only ice-free region of the Soviet Arctic. The Soviet Union must, therefore, be very careful how and under what conditions Western firms operate in these waters. Soviet Energy planners surely must rigorously justify to their military counterparts any foreign involvement in the region.

While the ice-bound waters of the Sea of Okhotsk are not as strategically important as the Barents Sea, they are nonetheless bordering on Japan which the Soviet Union perceives as a staunch American ally. Considerable caution must be exercised in dealing with Japanese firms in this region. The Caspian Sea of all the major offshore regions is probably the least problematic from a security point of view. It is the region which seems least dependent upon foreign assistance.

Given these considerations, the Soviet Union has been careful about its negotiations with foreign firms. It has in each case made sure that before a deal was struck the foreign partner had as much or more at stake than the Soviet Union.

In the case of Japan, the Soviet Union felt safe to enlist Japan's support given Japan's dependency upon foreign sources of energy. The Soviet Union could, therefore, push a harder bargain, confident of Japanese commitments. The fact that operations between the two countries have scaled down considerably since the fall in world oil prices, implies that the Japanese are in a less-desperate position. As a result the Soviet Union may not feel it can be the senior partner in the negotiations.

Similarly with Norway and Finland the Soviet Union has always attempted to share risk. Norway has much at stake as a member of NATO and the Soviet Union knows it. By expanding its offshore industry Norway encroaches upon disputed territory with the Soviet Union in the western portion of the Barents Sea. The Norwegians will have to exercise caution then so as to not appear hostile. The Soviet Union will surely use this as leverage in negotiations over the sale and use of exploration equipment in the region. Finland, with the largest portion of its trade tied to the Soviet Union, is committed to compromise in the sale of offshore equipment, and has been the best supplier of both equipment and expertise to date.

How are the effects of overriding political conflicts between the East and West related to the use of and assistance from Foreign partners?

The effects of East-West tension in the offshore petroleum industry have largely been manifested in American equipment embargoes. Two situations have adversely affected the offshore industry. The first was the Carter administration's sanctions levied over the Soviet invasion of Afghanistan; the second was the Reagan administration's sanctions over the Soviet-Western European gas pipeline.

The Carter sanctions involved the cancelling of hundreds of millions of dollars of equipment in leased parts and whole rigs the Japanese were prepared to sell or lease to the

Soviet Union. The $450 million McDermott Inc. deal (see Table 5.2), was suspended under these sanctions. The effect this had was virtually to stop all essential equipment deliveries to the Soviet Union by its Japanese partner. It made the Soviet Union realize how dependent it was on American technology in its dealings with Japan. The comprehensiveness of American technology became even more evident when the United States Congress halted equipment sales of licensed American parts to Schlumberger (France) and other European firms who were only indirectly involved in the development of the Sea of Okhotsk with the Soviet Union.

The Soviet Union, however, must have been pleased by the fact it was involved with a partner as concerned with the Carter sanctions as itself. Japan lobbied early in 1980 to the USDC to lift the embargoes, thereby proving to the Soviet Union Japan's commitment to the project. It was surely a benefit to the Soviet Union that one of America's strongest allies was arguing to the benefit of the Soviet Union. The Japanese were successful in having Congress lift the equipment embargoes of those firms involved in SODECO.

However, the Soviet Union was always uneasy about the dependency of Japanese firms on American technology, and continually tried to involve a number of countries having their own offshore technology. When the Reagan administration extended in 1982 the Soviet gas pipeline sanctions to all American energy equipment being sold and licensed to the Soviet Union, this confirmed for the Soviet Union the difficulty of dealing with American allies. Efforts to enlist non-American suppliers were therefore stepped up and were considerably successful. Finland, Norway, France, West Germany and the United Kingdom all became involved in equipment sales of offshore technology to the Soviet Union in spite of the American embargoes. In this, the offshore industry benefited largely because of the individual experience of Norway, Finland and the United Kingdom in offshore petroleum, and the independent line usually taken by France in these East-West conflicts. The Soviet Union was engaging foreign partners less dependent upon the Americans for offshore equipment.

The general dissent among American allies over the gas pipeline sanctions and its coverage of all oil and gas equipment was to the benefit of the Soviet Union. In July of 1982, the EEC began to lobby the United States to lift the general sanctions. A further estimated billion dollar loss in oil and gas equipment sales by American firms because of the sanctions forced the Reagan administration to lift the sanctions in November 1982.[11] The lifting of sanctions did considerably speed-up the Sea of Okhotsk exploration and development. But the Soviet Union could no longer believe in the sanctity of contract when American firms were involved. That is why in the development of the Arctic waters, and increasingly in the Caspian Sea, non-Americans are now being enlisted. The non-American contacts the Soviet Union forged during the Reagan sanctions have become lasting commitments to the benefit of the offshore petroleum industry in the Soviet Union.

The fact that the Soviet Union did not severely limit its activity after the setbacks of the Reagan administration sanctions, but went ahead both with purchases of equipment and plans to upgrade offshore construction yards at Baku and Vyborg, shows that it is serious about the industry's development. What the Soviet Union did, and is continuing to do, is enlist the assistance of those countries with an interest in the development of the Soviet petroleum industry.

The ability to increase imports of the necessary technology from non-American sources will probably mean some form of cooperation will resume between Japan and the Soviet Union in the Sea of Okhotsk. The region is too well explored. Both nations have too great a stake in its development for the area to be left alone. The Soviet Union will probably gain the necessary ice-resistant rigging and with Finnish help, will improve its drillship capacity, in order to negotiate with Japan.

The Arctic seas, particularly the western Barents, will be an area of continued success and will probably lead to the most innovations in offshore technology. The sheer experience of the Norwegians and the continued interest shown to date will be the contributing factor in this. As the Soviet Union gains more experience in the Caspian Sea with deep-water drilling and production, it will be able to apply this expertise in the Arctic fields.

What are the systemic problems that inhibit the Soviet system's dealing with foreign firms and what have they done to alleviate these problems?

The first problem which inhibits trade, not only in the offshore petroleum industry but also in the whole system of foreign trade for the Soviet Union, is the effect of material balances upon trade negotiations. Since plans are so taut and there is little flexibility in response to demands on its resources, the Soviet Union prefers bilateral arrangements. This involves longer-than-normal negotiations (if Western firms are involved) in order for the Soviet Union to account for resource sales and equipment purchases in its material balance. Typical bilateral negotiations between the Soviet Union and its CMEA partners are very long sessions. Neither side expects the *valiuta* rubles to reflect the importance, priority, or relative scarcity of the items being negotiated. This is a reflection of the non-scarcity prices in the socialist economies and the fact that currencies are not convertible. One of the effects of nonconvertibility is that one does not gain access to Soviet goods by owning rubles. Such a situation would be too upsetting to the material allocation system in the Soviet Union. This has caused problems for Soviet foreign trade organizations (FTOs) when they deal with Western firms. Many western firms insist on a given level of Soviet purchases as a *quid pro quo* for allowing the Soviet Union access to its products. This constrains the FTOs since they may wish to simply buy in the cheapest market with hard currency. Further, due to the control over foreign trade by the Ministry of Foreign Trade, enterprises are deprived of the ability to specialize intelligently in trade with foreign partners.[12]

As a result the Soviet Union loses. It can not take full advantage of potential gains from trade. The foreign partner also loses. The partner is often subject to long negotiations, and to buying goods that cannot be sold (or even dumped) on Western markets. The fact that Japan held on for so long in its negotiations over the development of offshore petroleum in the Sea of Okhotsk was largely due to the continued high price for petroleum. That is, it still paid for the Japanese to carry on lengthy negotiations while the price for oil was still quite high in the mid-1970s. The fact the Japanese have expressed little interest in further development in the Sea of Okhotsk since 1984 is primarily due to the lower price for oil. This has hurt the exploration and development plans for the whole region. Norway is in a situation where it cannot help but cooperate. Its interests in the Barents Sea insure the Soviet Union a partner in this region, perhaps even when it is uneconomic for the Norwegians to operate in this zone.

The offshore industry was one of the first areas of the economy the Soviet Union saw that some special arrangement would have to be made. Joint venture legislation, as noted above, was passed as early as 1982 in this sector specifically to encourage the influx of foreign capital and technology. This early legislation, however, was not nearly as comprehensive as the current joint-venture legislation under Gorbachev, aimed primarily at raising the qualitative level of Soviet products in export through the acquisition of Western technology, and to gain access to Western managerial techniques. According to one source the joint-venture regulations are to:

i) create conditions to encourage foreign investors by including special low taxation rates and the right to freely repatriate profits and other assets belonging to the foreign partner;

ii) allow management decisions to be taken on a joint basis with each partner having a veto;

iii) encourage Western quality control;

iv) exempt joint ventures from plan targets;

v) make hard currency funds available for imports to enterprises that guarantee repayment in hard currency from exports within four years; and,

vi) allow enterprises to have direct relations with foreign firms and to

> *receive hard currency allocations as a reward for over-plan exports.*[13]

The relevance of this legislation to offshore petroleum should be viewed cautiously. The extent to which foreign firms such as Norway or West Germany are to be allowed equal decision-making authority in a region such as the western Barents is probably minimal. The ability of joint cooperation to yield petroleum within four years of exploration and drilling is doubtful. But the legislation is important in that it may allow for accommodation in other related sectors to offshore petroleum, such as in onshore Arctic resource development or in some related industry such as shipbuilding.

While these propositions may not seem workable at present, they surely are an improvement over long-stalled negotiations (as was the case with Japan and Gulf Oil Co.), time-consuming technical cooperation (the case with Norway), and costly buying with no accompanying expertise (the case with France, Finland and Japan). These have all characterized involvement with foreign firms in the past. Essentially, therefore, the extent to which the offshore petroleum industry will improve in its ability to attract foreign expertise is very much a function of the success of Gorbachev's entire economic reform.

Finally how has the Soviet Union dealt with or diffused the boundary disputes that have arisen over its exploration of the offshore waters?

There have been three areas where the Soviet Union has had to deal with territorial problems. The first is the Sea of Okhotsk with Japan, the second is the Barents Sea with Norway, and the third is the Baltic Sea with Sweden.

The dispute with Japan began in 1875 when Japan ceded Sakhalin to Russia in exchange for the Kuril islands. Japan regained control of the area on Sakhalin below 50 degrees north (called Karafuto) in 1905 after the Russo-Japanese War. However, in 1945 upon the defeat of the Japanese in World War II, the Soviet Union seized all of Sakhalin and the Kuril islands. While Japan renounced claim to part of

Sakhalin in 1951, it has not settled the Kuril islands yet. Because Japan lays claim to the islands, it will not settle on a marine boundary between the two island groups. The Norwegian dispute is in the western Barents Sea in an area known as the "Grey Zone". This zone was decided upon for fishing purposes between the two countries, however, the agreement in no way prejudices either side's claim to territorial rights. The boundary, as it pertains to mineral extraction rights, is very much under contention. Also, the shelf around Spitsbergen island is under negotiation as both countries exercise their rights to mine the island. The dispute with Sweden over oil on and around Gotland island is innocuous but nonetheless arose over Soviet activities in the Baltic Sea. The Soviet Union lays claim to half the island and the waters up to it, while Sweden claims the distance halfway between itself and the Soviet Union.[14]

The Soviet Union has largely ignored the dispute with Sweden. This is probably because the stakes are not high enough in the region (i.e., little offshore oil seems to exist). But with both Norway and Japan the Soviet Union has actively sought to position itself for *de facto* recognition.

The Soviet Union has had plans to develop the Sakhalin basin for quite some time. The primary reason for the involvement of Japan was its proximity to the region, and hence interest, and its access to the most up-to-date technology via American licenses. But incidental to this cooperation was Japan's and the Soviet Union's agreement on where the Japanese and Soviet regions were. The Japanese, in a sense by negotiating what the terms of foreign specialists' participation would be in the Soviet zone of the Sea of Okhotsk, have legitimized the marine border of the two countries.

With regard to Norway, the Soviet Union is taking the initiative in exploration of the zone by having no less than five research vessels and two drillships operating in and around the disputed border. The cooperation of Sudoimport and the Norwegian consortium BOCONOR on the joint development of the western portion of the Barents Sea is of course primarily for the advancement of Soviet expertise in the Barents Sea, but, as with the Japanese in the Sea of Okhotsk, formal cooperation is a considerable step toward territorial recognition. The more active the Soviet Union is in the exploration of these waters the stronger its position becomes when the boundary is finally to be decided.

As Arctic and Far Eastern exploration for petroleum increases, tensions in other areas of unsettled boundaries--such as with Canada in the Laurentin basin and the outer Beaufort Sea, with the United States in the Sea of Chukotsk and the Bering Sea and with Denmark in the Angara basin, regarding Greenland--will increase. These disputes will probably take the form of mineral rights, which will eventually lead to some recognition of sovereignty. If so, then Soviet military and political planners may see an advantage to bolstering the offshore petroleum industry for reasons other than strictly energy development. Recognition (even *de facto*) of Soviet borders by its Far Eastern and Arctic neighbors would be of considerable advantage to the Soviet Union, much the same as was the recognition by West Europe, Canada and the United States of the borders of East Europe in 1975 with the signing of the Helsinki Accords.

Prospects

Certain things must be remembered in attempting to assess the future of exploration and development of the Soviet offshore waters. First, the immediate goal is to add to proven reserves and therefore halt the decline in the reserves-to-production ratio of the petroleum industry generally. The long-term goal is to ultimately produce petroleum from these regions. Second, realization of this is a very long process. As was mentioned earlier, it took the Norwegian industry together with cooperation from the international petroleum industry over ten years to acquire enough experience to operate and produce petroleum from a major field. The Soviet Union, with its tendency toward autarky in the energy sector, can be expected to take longer than this.

The Soviet Union would seem to have all the right factors in place to accomplish the first goal of adding to explored reserves. It has five world-class drillships, five large research vessels, twelve semisubmersible rigs, and has organized and managed to construct its own jack-up rigs, at least for temperate waters. Further, it has forged important foreign contacts, many of which have only limited if any dependence on the United States for their offshore technology requirements.

It would seem that the longer-term goal of production from the offshore waters also has good potential. The Soviet Union is making some important changes in its foreign trade sector, which if successfully implemented could lead to valuable collaboration in the offshore petroleum industry. Certainly the engagement of Norway, Finland and Japan as partners, each with an intrinsic interest in seeing a healthy Soviet offshore petroleum industry, is a good sign.

There is every indication the offshore petroleum industry will continue to attract foreign technology and increase in size. With predictions of a crude oil price per barrel of $20 plus by mid-to-late 1988, this can only make the industry's development all that more important. This will tend to increase the incentive for countries with limited petroleum supplies, such as France and Japan, to sweeten joint development plans. Increased petroleum demand will mean foreign consortia and individual companies will be able to rely on bank financing for the operations. The fact that American sanctions have been lifted and that the American domestic market for offshore technology is in serious trouble will mean that numerous American firms will try to gain access to the Soviet market. While the Soviet Union will be dubious of any ongoing partnerships that rely on continued deliveries of American equipment, the Soviet Union could stand to gain considerable technology benefits for as long as tensions are relaxed. This will supplement their new sources of technology from other countries.

Finally, the Soviet leadership seems committed to the industry's development. Although the proclamation in the 12th Five-Year Plan is the formal commitment, the concrete evidence of the present leadership's commitment to development is the planned purchase and production of over 50 mobile rigs [15], and an unknown number of research vessels, to explore and develop the offshore waters.

Notes

1. See Chapter 2, ff. 35.
2. This is the opinion of A.A. Makarov cited in Chapter 2, ff. 35.
3. Osnovnie napravleniia: ekonomicheskogo i sotsialinogo rasvitiia SSSR na 1986-1990 gody i na period do 2000 goda, (Moscow: Politisdat, 1986), p. 41.
4. See Business Eastern Europe (March 17, 1986), p. 81.
5. This is estimated very crudely as follows:
There are 21 technology imports for which the value is known. As these are fairly wide ranging in type they can be used as a proxy for most of the imports. The cumulative value of these imports is then added and divided by 21 and then multiplied by 79 (as there have been 79 imports since the first jack-up the "Khasar"). This gives a rough estimate of $9.383 billion.
6. See Table 4.1.
7. See Business Eastern Europe (March 17, 1986), p. 82.
8. This option was noted by a U.S. attorney John Connor. See Business Eastern Europe (Feb. 10, 1986), p. 42. Estimates on Soviet asset position in the West from Business Eastern Europe (March 17, 1986), p. 84.
9. See Business Eastern Europe (Sept. 22, 1986), p. 298.
10. See Business Eastern Europe (Sept. 22, 1986), p. 298.
11. See Oil and Gas Journal (Nov. 29, 1982), pp. 36-37.
12. See Alec Nove, The Soviet Economic System, (London: Allan and Unwin, 1986), pp. 285-288.
13. See Business Eastern Europe (Sept. 22, 1986), p. 298 and (March 17, 1986), p. 82.
14. See Noroil (April 1987), p. 34.
15. Noroil (Aug. 1986), p. 110.

Abbreviations, Acronyms and Definitions

bbl	billion barrels
bcm	billion cubic meters
bd	barrels a day
BOCONOR	A consortium formed in 1983 of seven Norwegian companies, including Norwegian Petroleum Consultants, joined together to negotiate development of the western portion of the Barents Sea with the Soviet Union.
BP	British Petroleum
BZGO	Baku Plant for Deepwater Foundations
CGE	Compagnie Generale Geophysique
CIA	Central Intelligence Agency
CMEA	Council for Mutual Economic Assistance
COCOM	Coordinating Committee on Multilateral Export Control (part of NATO)
CPSU	Communist Party of the Soviet Union
EEC	European Economic Community
ETPM	Entrepose GTM pour les Travaux Petroliers Maritime
FRG	Federal Republic of Germany (West Germany)
FTO	foreign trade organization
FYP	Five-Year Plan
GKNT	State Committee on Science and Technology (USSR)

Gosplan	Soviet State Planning Commission (or committee), attached to the Soviet government
IEA	International Energy Agency
mbd	million barrels a day
mcm	million cubic meters
mty	million tons per year
n.a.	not available
NATO	North Atlantic Treaty Organization
OECD	Organization for Economic Cooperation and Development
OPEC	Organization of Petroleum-Exporting Countries
OTA	Office of Technology Assessment
OTC	Offshore Technology Conference
SODECO	A consortium consisting of 17 Japanese companies created in 1974, including C. Itoh and Co., Marubeni Corp., and Japan Petroleum Development Corp., and formed to undertake exploration and development in the Sea of Okhotsk off the island of Sakhalin with the Soviet Union.
tbd	thousand barrels per day
toe	tons of oil equivalent
TsSU	Tsentral'noe statisticheskoe upravlenie (central statistical agency of the Soviet Union)
UIE	Union Industriel et l'Enterprise

U.K.	United Kingdom
USDC	United States Department of Commerce
U.S.	United States
WMP	world market price

Technical notes

1. I have used both million tons per year and thousand barrels a day in this study. The conversion ratio is 7.29 barrels per ton of standard crude oil. I have tried to keep all measures consistent with this conversion. The reader should keep in mind that there are other ratios depending upon the degree of refinement of the petroleum.

2. I have used the Library of Congress Transliteration system for all Russian words used in the text. A few place names, such as the Caspian Sea and the Baltic Sea, were exempted from this system because of their common usage. Some of the names used in the Caspian Sea section of the text come from Turkish via English language sources and have been copied directly.

3. All dollar values are in US dollars.

Bibliography

Asdov, R.I. *Gazovaia promyshlenost*, June, 1984, no. 6, pp. 44-50.

Baku VYSHKA. January, no. 1, in JPRS UEN-85-011, April, no. 4 in JPRS UEN-85-016, May, no. 5 in JPRS UEN-85-017, June, no. 6 in JPRS UEN-85-017, 1985.

Bergsen, H.O., A. Moe, and W. Ostreng. "In Search of Oil and Security: Soviet Interests in the Barents Sea," *Noroil*, August, 1986, pp. 109-115.

Business Eastern Europe. February 10, September 22, March 22, 1986; September 13, 1985; January 15, 1982; May 2, April 18, January 20, 1980; July 13, 1979.

Canadian Department of National Defence. *Soviet Offshore Oil and Gas Resources*. Ottawa. Unpublished paper. December, 1984.

Christian Science Monitor. November 21, 1979, p. 7.

CIA. *Prospects for Soviet Oil*. Washington: National Foreign Assessment Review Agency, April, 1977, ER-7710270.

____________. *The World Oil Market in the Years Ahead*. Washington: National Foreign Assessment Review Agency, August, 1979, ER-10327U.

CIA. *International Energy Statistical Review*. Washington: National Foreign Assesment Review Agency, March 29, 1987.

Clarke, James W. "Petroleum Geology of the Offshore Arctic." Unpublished paper. March 1988.

Dienes, Leslie, and Theodore Shabad. *The Soviet Energy System*. New York: John Wiley and Sons, 1979.

East-West Markets. August 7, September 18, 1978; April 19, 1976; May 6, October 21, 1974; November 28, 1973.

East-West Trade Council Newsletter. vol.3, no. 2, January 31, 1975; vol. 2, no. 20, December 20, 1974.

The Economist. October 22, 1977 p. 94.

Economist Intelligence Unit. *The Soviet Union: Energy Review*, 1983, no. 3.

____________. *The Soviet Union: Energy Review*, 1983. no. 4.

____________. *Quarterly Economic Review of USSR*, 1986. no. 1.

Egleograh, E. Onu and E.J. Peters. "The Importance of the USSR North in the Total Soviet and World Energy Supply System," *Conference on Energy in the USSR and Eastern Europe*, University of Alberta, (May 27-28, 1977).

Egyed, P. *Western Participation in the Development of Siberian Energy Development: Three Case Studies. East West Commercial Relations Series. no. 22*, Ottawa: Institute of Soviet and East European Studies, Carleton University, 1983.

Ekonomicheskaia gazeta. March, no. 12, p. 11, April no. 18, p. 1, 1986.

Erdol und Kohle. February, 1981, pp. 52-54.

Far Eastern Economic Review. July 23, 1982, p. 47.

Financial Times of London. October 22, 1986, p. 29;

February 17, 1982, p. 6; November 23, 1978, p. 5.

Foreign Broadcast Information Service (FBIS). March 8, 1985 no. 8.

Gazovaia promyshlenost. vol. 1, no. 7 July, 1985 in JPRS UEN-85-018.

Globe and Mail. June 27, 1985, p. B5.

Goldman, Marshall. *The Enigma of Soviet Petroleum: Half Empty or Half Full*. London: Allan and Unwin, 1980.

Gramberg, I.S. and I.V. Shkola. *Sovetskaia geologiia*, January, 1985, no. 1, in JPRS UES-85-006-66A.

Gustafson, Thane. "Soviet Oil Policy and Energy Politics: 1970-1985." United States, 1986. (Mimeographed.)

Hannigan, John B. *Oil and Gas Development in the Soviet North: Exploration, Production, Transportation.* Ottawa: Department of Indian and Northern Affairs and Northern Development, December 5, 1986.

__________. and Carl H. McMillan. *The Soviet West European Relationship: Implications of the Shift From Oil to Gas, East West Commercial Relations Series.* no. 20, Ottawa: Institute of Soviet and East European Studies, Carleton University, 1983.

Hewett, Edward. *Energy Economics and Foreign Policy in the Soviet Union.* Washington: Brookings Institute, 1984.

Izvestiia. April 15, p. 3, April 18, p. 6, 1984.

Khvedchuk, I. "Put' k morskim mestorozhdenii," *Pravda*, April 3, 1978, p. 1.

Lecky, James R., and J.R. Lee. "Siberian Oil," *Siberian Resource Development*, 1979, pp. 113-123.

Leningradskaia Pravda, March 5, 1985, p. 1.

Meyerhoff, A.A. *The USSR Northern and Far Eastern Coasts: Petroleum Geology and Technology, Mining Activities, and Environmental Factors.* Ottawa: Department of Indian and Northern Affairs, 1980.

Moscow Narodny Press Bulletin, November 21, 1984; September 12, 1979; April 12, 1978.

Muller, Friedman. "Implications for the West, Including Defence, of the CMEA Energy Prospects," *CMEA Energy: 1980-1990*, NATO Colloquium April 8-10, 1981.

Narodnoe khoziaistvo SSSR, 1984; 1981.

New York Times, October 13, 1977, p. 13.

Noroil, August, p. 31, November, p. 38, 1984.

Nove, Alec. *The Soviet Economic System.* London: Allan and Unwin, 1986.

OECD. *Foreign Trade by Commodities.* Paris: 1987, vol. 2.

Office of Technology Assessment. *Technology and Soviet Energy Availability.* Boulder, Colorado: Westview Press, 1982.

Offshore, April, May, 1986; May 1985; January 1984; June, August, 1983; April, June 5, 20, September, November, 1982; May 1980; June 1979; June 1977.

Ofusa, Junnosuke. "Japan and Soviets Agree on Joint Oil and Gas Project," *The New York Times*, November 25, 1976, p. 41.

Oil and Gas Journal, July 11, 1988; January 19, April 6, May 4, 1987; March 10, 21, 31, April 7, May 19, June 9, August 4, 11, 25, September 8, 1986; April 15, August 5, September 2, November 4, 1985; February 6, June 4, November 12, December 3, 10, 1984; January 24, February 14, 1983; November 29, 1982; November 10, 1980; April 26, June 14, August 2, 1976; November 3, December 22, 1975; June 3, 1974; December 22, 1972.

Osnovnye napravleniia: ekonomicheskogo i sotsial'nogo razvitiia SSSR na 1986-1990 gody i na period do 2000 goda. Moscow: Politizdat, 1986.

Perchik, A.I., *Ekonomika osvoyeniia morskikh mestorozhdenii nefti i gaza*, Moscow: Nedra, 1987.

Petroleum Economist, January, March, 1988; April, May, 1987; May, June, September, 1986; February, March, April, June, July, August, 1985; April, September, October, December, 1984; June, 1983; April, 1982; July, 1981; April, 1980; January, 1979; May, 1978; May, September, November, 1977.

Petroleum Intelligence Weekly, November 26, 1979, p. 12.

Platt's Oilgram News, June 8, 1979, p. 4.

Pravda, February 7, p. 1, September 5, p. 1, and 7 pp. 1-2, October 16, p. 5, 1985; May 15, p. 1, 1983.

Press, F. and R. Siever. *Earth.* San Francisco, California: W.H. Freeman and Co., 1974.

Riva, Joseph P. "Soviet Offshore Oil and Gas," *Soviet Oceans Development Committee on Commerce.* Washington: US Government Printing House, October 1976.

St. John, B. "Offshore Exploration Potential," *1985 Offshore Technology Conference*, Houston: May 6-9, 1985, vol. 1.

Sel'skaia zhizn', October 26, 1984, cited in JPRS UEN-85-001.

Shushkov, V. "Mutually Advantageous Cooperation Between the USSR and Japan," *Soviet Foreign Trade*, May 1975, pp. 22-25.

Sotsialisticheskaia industriia, October, 17, 1984, cited in JPRS UEN-85-001.

Sotsialisticheskaia industriia, May 17, 1985, cited in JPRS UEN-86-018.

Sotsialisticheskaia industriia, August 22, 1985, cited in JPRS UEN-86-007.

Soviet Business and Trade, March 9, 1985; July 9, December 23, 1984; June 29, August 7, 1982; June 21, September 7, 1981; March 12, 26, May 21, 1980; July 18, September 14, 1979; May 24, 1978; October 21, April 13, August 17, 1977; September 29, December 8, 1976; February 4, March 18, 1974.

Soviet Geography, vol. 26, March 1985; vol. 25, December, 1984

Sovietskaia rossiia, December 23, 1984, p. 6.

Stern, Jonathan P. "Soviet Oil and Gas Production Exports to the West: A Framework for Analysis and Forecasting," Joint Economic Committee, *Gorbachev's Economic Plans*. vol. 1. Washington: U.S. Government Printing House November 23, 1987, pp. 500-513.

Suliemanov, A.B., Kuliev, R.P.; Sarkisov, E.I.; and Karapetov, K.A. *Ekspluatatsiia morskikh neftegazovykh mestorozhdenii*, Moscow: Nedra, 1986.

Vneshniaia torgovlia SSSR, 1986; 1985.

Wilson, David. *Soviet Oil and Gas to 1990*. Economist Intelligence Unit Report no. 90 (1980).

World Oil, August 15, 1977, p. 130.

Index

Made in the USA
Las Vegas, NV
31 October 2024

10676788R00116